A phytogeographical study of the vascular plants of West Greenland (62°20'-74°00'N)

Bent Fredskild

Meddelelser om Grønland, Bioscience 45 · 1996

Contents

Abstract 3
1. Introduction 3
2. Study area 4
2.1. Topography 4
2.2 Geology 4
2.2.1. Pre-Quaternary 4
2.2.2. Quaternary 4
2.3. Soil 4
2.4. Permafrost 6
2.5. Lakes and homothermic springs 6
2.6. Climate 6
2.7. Vegetation 7
2.8. Holocene history of the flora 9
3. Exploration of the study area 10
4. Material and methods 11
5. Taxonomical considerations 11
6. Results and discussion 18
6.1. Distribution types 18
6.2. Species diversity in Greenland 25
6.3. Delimitation of floristic provinces and districts 26
7. Acknowledgements 26
8. References 29

Appendix (Maps 1-379) 31

Accepted: 1 December, 1995
ISSN 0106-1054
ISBN 87-90369-03-3

A phytogeographical study of the vascular plants of West Greenland (62°20'-74°00'N)

BENT FREDSKILD

Fredskild, B. 1995. A phytogeographical study of the vascular plants of West Greenland (62°20'-74°00'N). Meddr Grønland, Biosci. 45, 157 pp. Copenhagen 1996-00-00.

Phytogeographically and climatically West Greenland includes parts of the low arctic and high arctic areas. The present vegetation and a summary of its history since the last glaciation based on pollen- and macrofossil analyses are briefly described. The determination of more than 55,000 herbarium sheets of native phanerogams has been checked. Before preparing the dot maps it was necessary to decide at which level taxonomically difficult genera, e.g. *Antennaria*, *Draba*, *Poa*, *Puccinellia*, and *Stellaria* would be treated. For all of these the criteria used are given. Dot maps have been prepared for 379 taxa. These maps have been grouped into 11 West Greenland distribution types, which clearly correlate with T. W. Böcher's biological distribution types (Böcher 1963).

The collecting intensity in the large area is rather uneven. This is illustrated by a map giving the number of collections at the 305 localities from which 50 or more collections are at hand. Another map giving the number of taxa at 29 well-investigated localities illustrates the species diversity in the region. Most low arctic localities have more than 130 taxa, with 215 at the richest locality (Godhavn/Qeqertarsuaq on Disko), while the number at no high arctic locality exceeds 150. Reference to the corresponding numbers in other parts of Greenland is given.

Based on the dot maps a new delimitation of West Greenland in floristic provinces and districts is presented. It differs somewhat from that in Grønlands Flora/The Flora of Greenland (Böcher & al. 1957, and later editions). The major alteration is that the boundary between the low and high arctic phytogeographic provinces is now placed through north Disko – Nuussuaq. As a consequence of this the southernmost district in the high arctic is divided into an outer (NWso) and an inner province (NWsi). Minor alterations are suggested in the southern part of West Greenland.

Keywords: Greenland, flora, distribution maps, distribution types, floristic provinces

Bent Fredskild, Botanical Museum, University of Copenhagen, Gothersgade 130, DK-1123 Copenhagen K. Denmark.

1. Introduction

The first part of Greenland to be floristically investigated was the inhabited part, viz. West and South Greenland, of which the major part now can be considered fairly well explored. This is illustrated by the fact that more than one third of the sheets in the Greenland herbarium of the Botanical Museum, University of Copenhagen, originate from that part of the country dealt with here: West Greenland between 62°20'N and 74°00'N. The present paper is the third part of a phytogeographical investigation of Greenland, initiated at the establishment in 1962 of Grønlands Botaniske Undersøgelse (Greenland Botanical Survey), the main purpose of which was to carry out the systematic collecting of plants all over Greenland. The first part published was on South Greenland south of 62°20'N (Feilberg 1984), and the second on North Greenland north of 74°N (Bay 1992). The fourth and final part, E.Greenland between 62°20'N and 74°N, is under preparation partly by Geoffrey Halliday, University of Lancaster, partly by Christian Bay, Botanical Museum, University of Copenhagen. Once this is finished, the basis for a synoptic phytogeographical study of the Greenland vascular plants and for a new, revised "Flora of Greenland" will be at hand.

2. Study area

The study area West Greenland stretches from 62°20' to 74°00'N (Fig. 1). The western half of Greenland is here divided into: North Greenland, Northwest Greenland (the area between 74°N and Humboldt Gletscher at c.79°N), West Greenland (the present study area), and South Greenland (south of 62°20'N).

2.1. Topography

In the southern part of the area, between the huge glacier Frederikshåb Isblink (c. 62°30'N) and the ice cap at Maniitsoq (Sukkertoppen, 66°N), the topography is alpine, with fairly large areas above 1000 m a.s.l. Apart from the large peninsula northwest of Nuuk (Godthåb) there is only a narrow rim of coastal lowland (Fig. 2). Several nunataks penetrate the Inland Ice. The major part of the area between the Maniitsoq ice cap and the long fiord Kangerlussuaq (Søndre Strømfjord) is a highland, as is the area northeast of Sisimiut (Holsteinsborg). Apart from this the area between the head of Kangerlussuaq (67°N) and Disko Bugt (69°N) is a lowland, with rounded hills only exceptionally exceeding 400 m. The major part of the isle of Disko, of the large peninsulas Nuussuaq and Svartenhuk, and of the interjacent isles and peninsulas is a basaltic plateau landscape with steep slopes. North of Svartenhuk an archipelago with only narrow ice free rims on the mainland is intersected by the wide, ice-filled fiord Upernavik Isstrøm at 73°N.

2.2. Geology

2.2.1. Pre-Quaternary

The following summary description of the geology of the area is based on Escher & Watt (1976).

From the southern part of West Greenland to the highlands at Sisimiut and the inland south of Kangerlussuaq the bedrock consists of archaean gneisses, locally with archaean supracrustals like amphibolite or metasedimentary gneiss (Fig. 3). North of this area to c. 69°N on the mainland the Precambrian Nagssugtoqidian mobile belt is made up mainly of reworked older basement gneisses and granite, locally with interlayered and folded belts of metasediments and metavolcanics. In small areas of the southern Disko this basement is not covered by basalt. Another Precambrian mobile belt, the Rinkian, covers the eastern part of the area to 74°N. It can be divided into three parts, viz. from the south: a) an area of gneiss to c. 71°N with small areas of metasediments just south of the basis of Nuussuaq, followed by b) an area largely of metasediments northward to the base of Svartenhuk (c. 72°30'N), followed by c) an area of mainly granites. In the southern part of b) the Marmorilik Formation contains exposures of dolomitic and calcitic marble, with a layer of lead and zinc.

Sediments of mainly Cretaceous age cover the eastern part of Disko, part of the western half of Nuussuaq, and small areas of Svartenhuk. Over the major part of Disko, the western half of Nuussuaq, and Svartenhuk these sediments are covered by Tertiary basalts.

2.2.2. Quaternary

According to Funder (1989) all of West Greenland, excluding only some high mountains near the coast, was covered by the Inland Ice during the Late Wisconsinan Sisimiut glaciation. However, based on biological data on bryophytes Mogensen (1988) suggests that areas in the Disko-Svartenhuk area served as refugia during Pleistocene to the extent that this is reflected in the extant distribution pattern, and Ingolfsson & al. (1990) discuss evidence that some coastal, lowland slopes of west Disko were unglaciated, apart from local glacier tongues. Likewise, the isolated high altitude occurrences on Disko and Nuussuaq of some high arctic phanerogams can be taken as evidence of refugia here (distribution type 2a, see below).

2.3. Soil

In the areas of gneissic-granitic basement the dominating soil types are arctic brown soils (occasionally slightly podzolised), lithosols, and upland and meadow tundra soils. Analyses have been carried out, e.g., at Sisimiut and Ilulissat (Jakobshavn) (Stäblein 1977, Fredskild 1961), at three stations just south of the investigation area at c. 62°N (Kj. Hansen 1969), and on Tugtuligssuaq (c. 75°N) just north of the area (Jakobsen 1988). By far most measurements of pH ranges from 4 to 6. This is the case too in the gneissic upland around the western end of the long lake Tasersiaq (680 m a.s.l.) northeast of the ice cap at Maniitsoq. Here, the soil-forming processes of podzolisation is operative at much reduced intensity, yet

Fig. 1. Map of West Greenland showing meteorological stations and most other localities mentioned in the text. The insert map shows the boundaries between South, West, Northwest and North Greenland.

Fig. 2. Topographic map of West Greenland. On this and all other maps white areas, delimited by thin lines, indicate lakes, fiords, etc., whereas ice caps and the border of the Inland Ice are not marked by lines.

Fig. 3. Geologic map of West Greenland, modified from Escher & Watt (1976). Signatures: 1: basalt, 2: sediment, mainly of Cretaceous age, 3: metasediments, 4: gneisses. North of line A is continuous permafrost, between A and B it is discontinuous, and south of B only sporadic (after Weidich 1968).

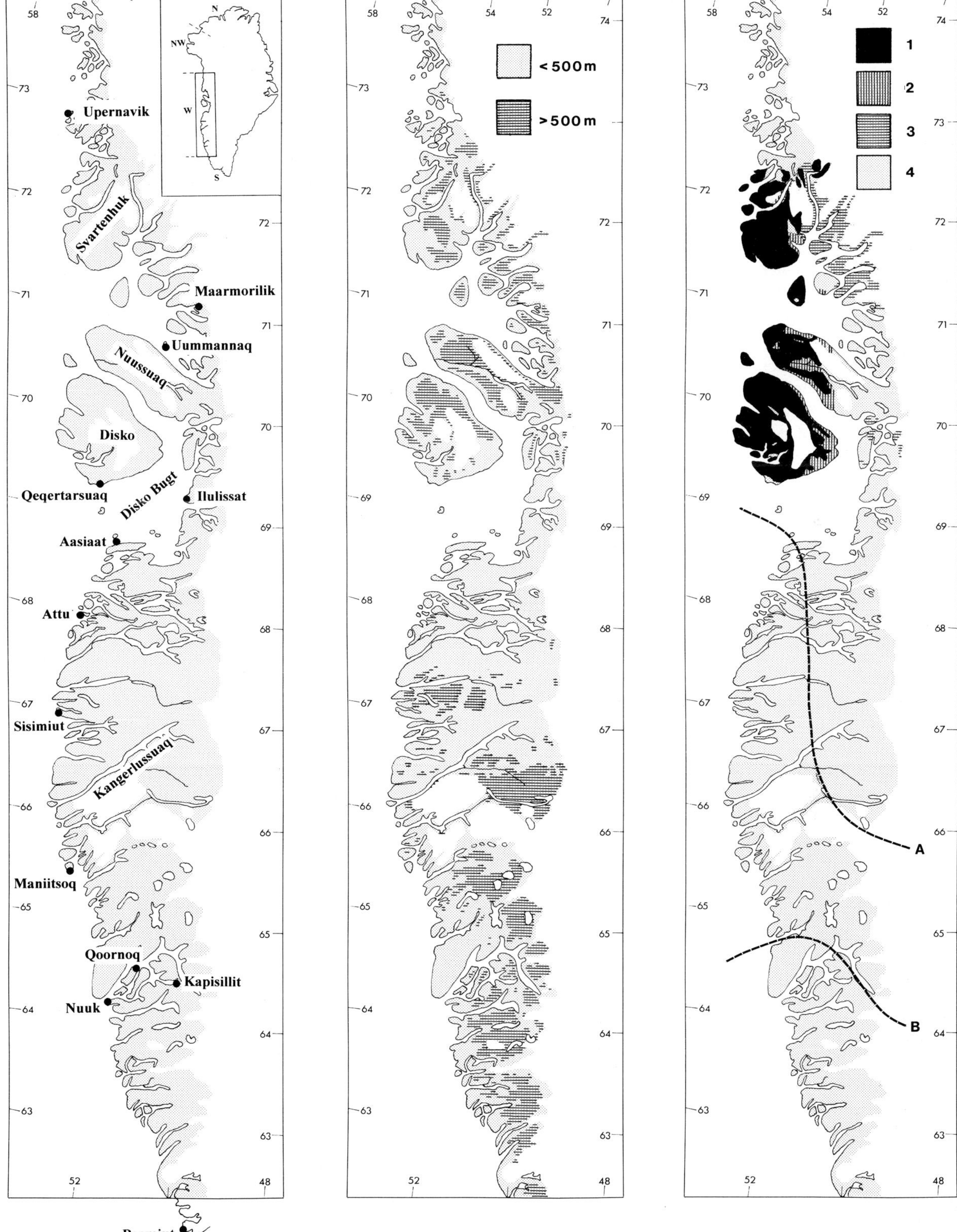

Fig. 1. Fig. 2. Fig. 3.

in favourable situations clearly recognizable podzolic soils are seen (Holowaychuk & Everett 1972). Below the Holocene upper marine limit, usually 50-100 m a.s.l., raised marine clay locally occur.

In the arid inland, especially around Kangerlussuaq, the soils often show layers of fine-grained, loess-like aeolian sediments, and pH ranges between 6 and 8. Here, acid soils only occur under mossy dwarf-shrub heaths on north-facing slopes. Salt efflorescences are frequent in depressions that are wet during spring, and around saline lakes (Böcher 1949, 1959, Dijkmans & Törnqvist 1991, Fredskild & Holt 1993, K. Hansen 1970). Measurements of conductivity and pH in connection with vegetation analyses are found also in Böcher (1954, 1963).

In the area of basalt and sedimentary bedrock most soils, apart from the A-horizon, are neutral-alkaline (e.g. Petersen 1981). Especially northwards, cryoturbation causes the formation of hummocky or patterned ground. Vestergaard (1978) describes the vegetation and soil conditions of salt-marshes on Disko and Nuussuaq.

2.4. Permafrost

South of 64°-65°N there is only sporadic permafrost (Fig. 3). North of this, discontinuous permafrost occurs southwest of a line from Aasiat (Egedesminde) to the inland at 66°N, whereas continuous permafrost is found north and east of this line (Weidick 1968).

2.5. Lakes and homothermic springs

Generally, the lakes and ponds of the coastal parts of the gneissic areas are oligotrophic and slightly acid, with conductivity well below 100 µS, in 182 of the 402 lakes measured even below 50 µS. Lakes of the interior are mesotrophic, with conductivity of the saline lakes around Kangerlussuaq exceeding 3000 µS (Böcher 1949, Fredskild 1992, K. Hansen 1967, Røen 1962). In the basaltic areas almost no permanent lakes are to be found. However, on the southern half of Disko many homothermic springs occur (Halliday & al. 1974). Most have temperatures of only 2-4°C, while a few exceeds 10° with 18.5 °C in the warmest. Because of a prolonged growing season which starts in cavities under the snow before general snow melt quite a number of southern species have their northernmost occurrences at the springs (Feilberg 1985). A few springs with temperatures only slightly above 0°C occur on Nuussuaq and Svartenhuk. Pingos and mudvolcanoes occur in the basaltic area, especially in the large, east-west running valley on Nuussuaq.

2.6. Climate

The climate of West Greenland varies from low arctic with mean temperatures in the south above 0°C for six months, to high arctic in the north with only four months above 0°C. Most meteorological stations are placed in the towns at the outer coast, having a cool, humid climate. Generally, winds from south and west prevail during summer, from north and east during winter. Inland, especially where coastal mountains or the ice cap at 66°N form an orographic barrier, the climate is continental. This is most pronounced at the head of the 175 km long Kangerlussuaq (Fig. 1), but also evident at the head of Godthåbsfjord east of Nuuk at 64°N, at the two long fiords at 68°N, and further to the north, in the fiord system at Uummannaq just northeast of Nuussuaq, c. 70°30'N .

Besides giving mean temperatures and precipitation at some West Greenland stations Table 1 also includes Conrad's continentality index (C) and de Martonne's humidity index (H) (Tuhkanen 1980). C is calculated as

Table 1. Climate data from West Greenland. Sources: upper part: Putnins (1971), lower part: Danish Meteorological Institute, extracted from the annual reports, published by the Ministry for Greenland (Anonymous 1979-91). NB: no observations from Sisimiut 1988.

			Mean temperature (°C)				Precipitation (mm)				
Locality	Latitude	Period	Year	Warmest month	Coldest month	June-August	Year	June-August	Humidity index	Conrad Index	Degree days above 0°C
Upernavik	72°47'	1931-56	-6.4	6.0	-19.6	4.9	186		52	30	498
Uummannaq	70°41'	1931-56	-4.0	7.8	-15.3	6.5	201		34	26	684
Qeqertarsuaq	69°15'	1931-56	-3.2	8.0	-13.9	6.8	391		58	24	720
Ilulissat	69°13'	1931-56	-3.9	8.2	-14.4	7.0	269		44	25	736
Attu	67°56'	1931-56	-3.8	6.7	-14.9	5.7					622
Nuuk	64°10'	1931-56	-0.7	7.6	- 7.7	6.7	515		55	13	809
Qoornoq	64°32'	1931-56	0.0	9.7	- 8.0	8.7	335		34	17	1051
Kapisillit	64°27'	1939-56	-0.7	10.9	- 9.6	9.7	255		27	22	1137
Aasiat	68°42'	1979-91	-5.2	6.2	-19.2	4.8	307	90	64	30	508
Kangerlussuaq	67°00'	1982-91	-6.2	10.6	-26.0	9.1	152	73	40	50	990
Sisimiut	66°55'	1979-91	-4.4	6.6	-18.5	5.3	369	134	66	30	597
Nuuk	64°10'	1979-91	-2.0	6.6	-11.0	5.4	662	220	83	17	613
Paamiut	62°00'	1979-91	-1.1	5.7	- 9.1	4.7	851	263	96	14	586

$$\frac{1.7A}{\sin(phi+10°)-14}$$

A is the annual temperature range in °C (warmest to coldest monthly mean) and phi is the latitude. Torshavn on the Faroe Islands has 0, Verkhoyansk in Siberia 100. H is calculated as

$$\frac{P}{T+10}$$

P is annual precipitation in mm, T is annual mean temperature in °C. Degree days are based on monthly means.

Table 1 has been split into two parts, covering different periods, as very few meteorological stations have long series of observations. However, when available such long series have shown a general decrease in mean temperature of 1-2°C from the first to the second period. This is best illustrated by the only station (Nuuk) common to both parts: the yearly mean temperature has decreased from -0.7° to -2.0°C, and the precipitation increased from 515 to 662 mm, thus giving an increase in C from 13 to 17 and in H from 55 to 83, and a drastic decrease in degree days from 809 to 613.

The increasing continentality with distance from the outer coast is clearly shown by the three stations along Godthåbsfjord (Nuuk, Qoornoq, Kapisillit, Fig. 1). The summer temperature increases from 6.7° to 9.7°C, precipitation decreases from 515 to 255, or, expressed in indices, C increases from 13 to 22, H decreases from 55 to 27.

All the year round, especially inland areas, are exposed to heavy, warm foehns from the inland ice. During the summer this may cause dust storms, and during winter temporary melting of the snow surface followed by freezing results in icing. At the head of Kangerlussuaq this, in combination with the very low precipitation, results in an extremely thin or even missing snow cover over large areas, exposing the vegetation to extreme conditions.

2.7. Vegetation

Warming (1888) gave the first detailed description of part of the West Greenland vegetation, and since then several papers, often including analyses, have dealt with the vegetation of larger or minor areas. The most detailed, including references to all earlier papers, are those of Böcher (1954, 1959, 1963), which describe the vegetation mainly around Maniitsoq, Sisimiut, the head of the fiords Kangerlussuaq and Arfersiorfik, and on south Disko. Besides giving many vegetation analyses he discusses the plant communities, their mutual relations, and their characteristic species. The species forming a plant community are ascribed to one or more of the following types: area-geographical differential species, climatic indicator species, ecological differential species, and ecogeographical guiding species (Böcher 1954:10), which formed the basis of eleven biological distributional types (Böcher 1963). In this paper he described the many types of plant communities, grouping them into: deciduous scrubs, dwarf-shrub vegetation, xerophilous grassland, meso-hygrophilous herbaceous communities, communities of bog and marsh plants, and communities of aquatic plants. Since that time only a few vegetation analyses have been published (Fredskild & Holt 1993, Philipp 1978, Vestergaard 1978).

The following summary descriptions, based on field notes and the yearly G.B.S. (Greenland Botanical Survey) reports, give an impression of the general change in vegetation from south to north, and differences from the inland to the outer coast.

At the head of the fiord systems at 64°N dwarf-shrub heaths are dominant, with *Betula nana* and Ericales each covering one half of the area. Among Ericales *Vaccinium uliginosum* ssp. *microphyllum* and *Ledum* are most important, *L. groenlandicum* in the lowland and in humid, mossy heaths, and *L. palustre* ssp. *decumbens* at higher elevations and in drier heaths. *Empetrum nigrum* ssp. *hermaphroditum* dominates the upland heaths, which include *Phyllodoce coerulea* and *Loiseleuria procumbens. Salix glauca* is a common heath plant. Fen-like communities are few, mainly restricted to lake shores or represented by hummocky communities with dwarf-shrubs on top of hummocks, and *Eriophorum angustifolium* ssp. *subarcticum*, *Scirpus caespitosus*, *Carex saxatilis*, and *C. rariflora* in between them. On moist south-facing slopes and along streams are open, 1-2 m high scrubs of *Salix glauca*, occasionally with many *Alnus crispa. Juniperus communis* and herbs like *Artemisia borealis*, *Saxifraga paniculata*, *Thymus praecox*, *Euphrasia frigida*, *Potentilla tridentata*, but only few grasses, are found on dry slopes. The lower zone of the tiny salt-marshes are dominated by *Puccinellia phryganodes*, while in the upper zone grow *Stellaria humifusa*, *Carex glareosa*, and *Potentilla egedii*.

Further out in the fiords *Empetrum* gradually becomes the dominating heath plant, and *Betula nana*, still fairly common, is restricted to the drier heaths. *Vaccinium uliginosum*, mostly sterile, and *Ledum* spp. are less frequent. *Salix herbacea* dominates snow-patches and late, snow-protected heaths. Herb-slopes with e.g. *Taraxacum croceum*, *Potentilla crantzii*, *Sibbaldia procumbens*, and *Veronica alpina* are seen. *Alnus* disappears as do the *Salix glauca* scrubs. The widespread, usually hummocky fens are dominated by *Salix arctophila*, which grows on the sides of the hummocks, and *Scirpus caespitosus*, *Carex bigelowii*, and *Eriophorum angustifolium* in between. At the outer coast *Empetrum*-lichen heaths are dominant, and *Salix herbacea-Harrimanella hypnoides* snow-patches with *Loiseleuria procumbens* are frequent, as are fen-like communities. The rudimentary *Salix glauca* scrubs are only knee-deep.

At the head of Kangerlussuaq (67°N) dwarf-shrub heaths are dominant. On south-facing slopes and dry, level ground are *Betula nana-Vaccinium uliginosum* heaths containing *Kobresia myosuroides*; on north-facing slopes mossy *Betula nana-Ledum palustris* ssp. *decumbens* heaths contain *Vaccinium uliginosum*, *Pyrola grandiflora*, and *Cassiope tetragona,* locally with the grasses *Calamagrostis lapponica* and *Poa pratensis*. Very dry, open heaths, dominated by *Betula nana*, *Dryas integrifolia*, and *Carex nardina* are found on top of low ridges. On thin, loess-like soil on dry south-facing slopes a steppe-like vegetation is dominated by *Kobresia myosuroides* and *Carex supina* ssp. *spaniocarpa*. Here, *Calamagrostis purpurascens* and *Poa glauca* are frequent, while *Festuca brachyphylla*, *Hierochloë alpina*, *Elymus violaceus*, *Carex nardina*, and the forbs *Potentilla hookeriana*, *Artemisia borealis*, *Campanula gieseckiana*, *Melandrium affine*, and *M. triflorum* are sporadic. Fens found mainly along the shores of lakes and ponds, are dominated by *Eriophorum angustifolium*, *E. scheuchzeri*, *Carex rariflora*, and *Calamagrostis neglecta*, with *Juncus castaneus* and *J. triglumis*. A fen-like grassland, which dries out during the summer, is dominated by *Calamagrostis lapponica*. A few *Salix* copses are found along streams, or lower and more open copses on deep soil at the foot of south-facing slopes. Snow-patch and herb-slope vegetations are all missing. Of special interest is the vegetation of dry or drying-out, alkaline soil, often with salt efflorescences, which support *Lomatogonium rotatum*, *Gentiana detonsa*, *Primula stricta*, *Braya linearis*, *B. novae-angliae*, *Juncus castaneus*, *J. arcticus*, *Puccinellia deschampsioides*, and *Carex boecheriana*. Lakes and ponds in alkaline areas are rich, both as regards vegetation and conductivity. Due to the physical conditions salt-marshes are very rare.

At the head of the fiords northeast of Sisimiut the lowland vegetation very much resembles that at the head of Kangerlussuaq. Above 300 m a.s.l. small snow-patches with *Salix herbacea*, *Harrimanella hypnoides*, *Gnaphalium supinum*, and *Antennaria canescens* can be found on north-facing slopes, and on protected sites occur fragmentary herb-slopes with *Erigeron humilis*, *Antennaria glabrata*, *Sibbaldia procumbens*, and *Potentilla crantzii*. Salt-marshes are well developed and include *Puccinellia phryganodes*, *Carex subspathacea*, *C. glareosa*, *Stellaria humifusa*, *Potentilla egedii*, *Triglochin palustre*, and *Cochlearia groenlandica*. At Sisimiut in the outer coast lowland herb-slopes and snow-patches are frequent. Another snowprotected, common vegetation type is dominated by *Luzula confusa* and *Empetrum hermaphroditum*, with abundant *Huperzia selago*. *Empetrum* dominate the many heaths.

Between the southeast corner of Disko Bugt and the Inland Ice (c. 68°30'N) the vegetation still reflects the fairly continental climate. *Betula nana* dominates the widespread lowland heaths, which at more humid sites include *Ledum palustre* ssp. *decumbens*, *Pedicularis labradorica*, and *Calamagrostis lapponica*. *Empetrum-Vaccinium uliginosum* heaths are fairly frequent, and on north-facing slopes, especially at a little higher elevation, *Cassiope-Phyllodoce* heaths occur. As everywhere in low arctic West Greenland *Salix glauca* is an important heath plant. On windswept sites *Dryas integrifolia* heaths are found, and on dry south-facing slopes steppe-like *Carex supina* communities are seen. The few *Salix* scrubs are 1-1½ m high. Snow-patches and herb-slopes are few and found only above 3-400 m a.s.l. The ponds are rich, e.g., with *Utricularia intermedia* collected in flower, which has been seen only three times in Greenland. The salt-marshes are well developed as they are all around Disko Bugt, with *Puccinellia phryganodes* and *Carex subspathacea* dominating the lower zone, *C. glareosa* the upper, which also includes *Stellaria humifusa*, *Potentilla egedii*, *Carex ursina*, and, near the upper transition, often *Mertensia maritima* and *Cochlearia groenlandica*.

This continental belt is narrow, the major part of the south coast of Disko Bugt being an outer coast lowland. Thus, the isles at Aasiaat are mostly covered by *Empetrum* heaths, rich in lichens and mosses, between the gneissic rocks, which are all covered by lichens. *Salix glauca* and especially *Betula nana* are infrequent, whereas *Salix arctophila* is common in the heaths. On north-facing slopes even down to sea level *Cassiope* heaths are common, i.a. with *Salix arctica*. *Salix herbacea* is very abundant in many communities, incl. also the true *S. herbacea-Harrimanella* snow-patches. Herb-slopes are few and small, all situated on south- or west-facing slopes at lakes. The fens are dominated by *Eriophorum angustifolium* (mostly sterile) and *Carex rariflora*. Most lakes and ponds are without phanerogams.

The south coast of Disko at Qeqertarsuaq (69°15'N) exhibits the most diverse flora and vegetation of low arctic Greenland, including luxuriant *Alchemilla glomerulans* dominated vegetation, with many low arctic species like *Leucorchis albida*, *Platanthera hyperborea*, *Listera cordata*, and *Epilobium hornemannii*, willow scrubs with *Angelica archangelica* ssp. *norvegica* at the homothermic springs, rich heaths and snow-patches, *Cassiope* heaths, and many types of fell-field vegetation, with many high arctic taxa. For further details see M. P. Porsild (1920), Böcher (1959, 1963), Philipp (1978), and Feilberg (1985).

On the south side of the base of Nuussuaq the vegetation at Qeqertap ilua (70°N, 51°W) still bear the imprint of low arctic, continental conditions, but it is somewhat impoverished. In the lowland *Betula nana* is the most common heath plant, followed by *Vaccinium uliginosum*, *Ledum palustre* ssp. *decumbens*, *Empetrum*, and, in less dry heaths, *Salix glauca*. Above 500 m these heaths are replaced by *Cassiope* heaths or *Loiseleuria-Salix herbacea* "heaths", with *Phyllodoce* and *Silene acaulis*, and by fell-fields. Herb-slopes are few, and no *Salix* copses are seen. On the contrary, many types of snow-patches, ranging from *Anthelia* dominated solifluction soil to *Salix herbacea-Stereocaulon canescens* vegetation, with *Antennaria canescens*, *Polygonum viviparum*, *Luzula spicata*, *Trisetum spicatum*, and *Agrostis mertensii*, are frequent. Fens are few, whereas frost-boil vegetation, either with i.a. *Sagina intermedia* and *Juncus biglumis*, or with *Tofieldia pusilla* and tiny specimens from the surrounding heaths vegetations, are fairly frequent on shallow ground.

Further to the west, on the northeast coast of Disko around Asuk (70°N, 53°W) the vegetation is high arctic continental. *Dryas integrifolia* heaths, *Dryas-Carex rupestris* heaths, and *Dryas-Salix arctica* heaths are dominant, whereas *Cassiope tetragona* only occurs in the gorges in certain belts, which are clearly related to the duration of the snow cover. *Vaccinium uliginosum* is fairly frequent, *Betula nana* rare. On dry sites *Carex nardina* and *Kobresia myosuroides* are common, sometimes accompanied by *Potentilla pulchella* or *Lesquerella arctica*. Snow-patches and frost-boil vegetation are sparse and small, and fens, ponds and, with one tiny, species-poor exception, herb-slopes are absent.

Further west, on the northwest coast of Disko, the vegetation is high arctic maritime. Mossy *Cassiope tetragona-Salix arctica* heaths cover vast areas, and solifluction soil and frost-boils, with open vegetation dominated by *Juncus biglumis, Polygonum viviparum,* and *Equisetum arvense*, are frequent. *Salix arctica* heaths with some yet always sterile *Vaccinium uliginosum* are frequent on southwest-facing slopes. *Empetrum hermaphroditum* only occurs at very protected sites. No *Betula nana* and only once was *Salix glauca* found. Only exceptionally in dry slope vegetations are seen *Carex nardina, C. glacialis, Potentilla vahliana*, and *Antennaria ekmaniana*. Only a few areas of species-poor snow-patch and herb-slope-like vegetation are seen. *Carex stans* fens occur along the rivulets, but neither *Eriophorum* fens nor ponds were seen.

So far, G.B.S. has not worked between Nuussuaq and Melville Bugt, and only few descriptions, in general terms and without analyses, are available. The following survey of the vegetation from 72°-73°N is based on Sørensen (1943). Heaths and fell-fields dominate. Heaths with *Empetrum hermaphroditum*, *Cassiope tetragona*, *Dryas integrifolia*, and *Vaccinium uliginosum* are common, with *Empetrum* being replaced by *Cassiope* northwards. In the heaths *Salix glauca* is northwards gradually replaced by *S. arctica*. *Dryas* heaths occur on drier sites. Other common heath types are dominated by *Cassiope tetragona*, and, when the duration of snow cover is too long for this species, by *Salix arctica.* In very open heaths of this type *Luzula confusa* is common. Herb-slopes and snow-patches with herbs are few and poor in species. Fens with *Eriophorum angustifolium*, *E. scheuchzeri*, and *Carex rariflora* sometimes include *Alopecurus alpinus*.

At the head of Laksefjord (72°30'N, 54°30'W) on the northwest side of the base of Svartenhuk the vegetation at protected sites resemble that at lower latitudes. Thus, the northernmost *Salix glauca* copse, 2 m high, is found here (M. P. Porsild 1912), and in rich heaths *Betula nana*, *Ledum palustre* ssp. *decumbens*, and *Phyllodoce coerulea* are dominant. On the east side of the base of Svartenhuk, an almost $2^1/_2$ m high *Salix* copse is seen at the head of the Uvkussigsat fiord (72°15'N, 53°45'W; K. Jakobsen, field notes).

2.8. Holocene history of the flora

The history of the flora since the last glaciation is fairly well known for the area northwards to Disko Bugt. Thus, pollen diagrams, and often also macrofossil diagrams for all or the major part of the Holocene have been published for four lakes in the Paamiut area (Kelly & Funder 1974), four at Godthåbsfjord (Fredskild 1973, 1984a), and one on the isle of Qeqertasussuk in the southeast corner of Disko Bugt (Böcher & Fredskild 1993). To these are added analyses of many peat deposits, mostly included in the papers mentioned, and preliminary analyses of other lake cores. Most important is the oldest Greenland core so far known, viz. from the outer coast isle Qeqertarsuatsiaq (68°26'N, 52°57'W). The basal 4 cm of gyttja have been dated at 11,320 +/- 140 B.P. (K-5133). This and all dates mentioned here are uncalibrated ^{14}C years. Pollen and macrofossil diagrams have been prepared for this lake (Fredskild unpubl.). So far, only late Holocene diagrams are available from the continental inland around Kangerlussuaq (Eisner & al. 1995). No long pollen diagrams are available from the Disko-Nuussuaq-Svartenhuk area, the first high arctic ones being from two lakes on Tugtuligssuak (75°22'N) in the Melville Bugt (Fredskild 1985).

The vegetational development in low arctic West Greenland can be summarized as follows:

Following the withdrawal of the ice, widespread, ubiquitous arctic pioneer plants spread over the raw, minerogenous soil. Dominating in the pollen spectra are *Oxyria digyna*, Poaceae, Cyperaceae, *Saxifraga oppositifolia* type, *S. caespitosa* type, *S. nivalis* type. *Minuartia/Silene* type, *Cerastium/Stellaria* type, Brassicaceae, *Chamaenerion*, and occassionally *Koenigia islandica*, *Sedum/Rhodiola,* and *Campanula*. Present also are seeds or fruits of, e.g., *Minuartia rubella*, *Silene acaulis*, *Carex bigelowii* type, *Carex nardina*, and *Chamaenerion latifolium*. Today only three limnophytes: *Ranunculus confervoides*, *R. hyperboreus* and *Hippuris vulgaris* grow in North Greenland. Fruits or pollen of these in the basal sediments show that they almost immediately immigrated to the region. There is still some uncertainty as to whether the ericaceous dwarf-shrubs were present at the beginning of deglaciation since mostly one or a few pollen are found in the very deepest sample(s), which, quite naturally, are very poor in contemporaneous pollen but contain comparatively many exotics, e.g. *Pinus*, *Pi-*

cea, *Ambrosia*, and *Alnus*, either blown in from long distance or rebedded from interglacial deposits. Long-distance dispersal or rebedding may also account for the ericaceous pollen. However, usually after 2-3 centuries these dwarf-shrubs had spread, and especially leaves and achenes of *Empetrum nigrum* ssp. *hermaphroditum* are frequently found.

No low arctic or boreal plants have been found in the oldest samples. However, by 9000 years ago pollen of e.g. *Angelica archangelica* and *Plantago maritima*, and seeds of *Galium brandegei* indicate a favourable climate, and around 8000 B.P. the climate became warmer than today. Among the shrubs *Salix glauca* reached the Paamiut area around 9000, Godthåbsfjord and Qeqertarsuatsiaq around 8000; likewise *Juniperus communis* ssp. *alpina* first reached Paamiut c. 7000 and did not appear in Godthåbsfjord until almost a millennium later. *Betula nana* spread from Iceland to East Greenland by 8000, but not until 6500 B.P. did it cross the Inland Ice to reach the interior Godthåbsfjord, from where it spread to the outer coast and to Disko in the following millennium. The American *Betula glandulosa* first reached Southwest Greenland at c. 5700, and later it spread to South Greenland. *Alnus crispa* was the last shrub to arrive; this happened c. 4000 B.P., by chance at the onset of a general cooling, traced all over West Greenland and also recorded in the Ice Cap cores. As a result of this climatic change ericaceous heaths expanded, and the abundance of warmth-demanding plants like *Juniperus*, *Betula nana*, and *Rumex acetosella* decreased.

Generally, limnophytes do not follow the same trend, since their distribution besides being dependent on temperature also relates to the trophic state of the water (Fredskild 1992). In early post-glacial time all lakes were rich in electrolytes, and mesotrophic species or those preferring/tolerating high content of electrolytes dominated. By 9000 B.P. the boreal *Potamogeton filiformis* and *Myriophyllum spicatum* ssp. *exalbescens* were in flower in eight of the analyzed Godthåbsfjord lakes. Gradually the water became more oligotrophic, *Isoëtes echinospora* immigrated, and *Myriophyllum spicatum* was replaced by *M. alterniflorum* ssp. *muricata*. Today *M. spicatum* does not grow in any of the eight lakes. The temperature decrease at 4000 B.P. added to the effect of the increasingly nutrient-poor water. Consequently, towards their northern limit, and in cool, coastal areas, most limnophytes today are sterile.

The discussion of which species, if any, had "wintered" the glaciation(s) on refugia in Greenland has been running for more than a century since the quarrels between Warming and Nathorst in the 1880s. In the 1930s T. Sørensen, P. Gelting, G. Seidenfaden, and T. W. Böcher continued the discussion in many papers (summarized in Böcher 1951). Until the first pollen diagrams were published (Iversen 1954) the discussion was only based on distribution types, but since then palaeobotanical evidence has added many facts. When considering also the temperature measurements from the Ice Cap cores it seems most likely that on any non-glaciated, preferably lowland site widespread or high arctic taxa of pioneer plants, e.g. *Saxifraga*, *Papaver*, Brassicaceae, Caryophyllaceae, and some Poaceae, and Cyperaceae may well have survived, and possibly also occassionally some Ericales.

Many authors quite naturally have seen the isolated occurrence of many high arctic plant species on north Disko and west Nuussuaq (distribution type 2a, see below) as a proof of a refugium here (e.g. Hultén 1937, Fig. 43). Most of these are absent in the area between Nuussuaq and the Thule area in NW.Greenland (76°-77°N). However, the whole area between Svartenhuk (Fig. 3) and Thule is gneissic, with one, tiny exception, so it might be argued that the disjunct distribution is edaphically conditioned rather than historically. The only lake found in the basaltic area that seemed suitable for palaeobotanical investigation (the isle of Hareøen northwest of Disko) was formed as a result of a giant earth slope only 4430 ± 85 years ago (K-4137, Fredskild unpubl.). Thus, the only hope of getting the ultimate answer to the question seems to be buried riverine or estuarine sediments with plant remains. Still, some odd distributions are not easily explained. For example, two isolated finds of *Braya linearis* in the eastern part of Disko Bugt (Map 347) are from moraines formed during the late 19th century ice advance, and the even more isolated occurrence of *Agrostis stolonifera* (Map 273) is from a recent alluvial fan just in front of the margin of the Inland Ice. The nearest known site of this species is at 61°15'N in South Greenland.

3. Exploration of the study area

The oldest plant collections from Greenland, gathered in a book herbarium dated 1739, were made by Paul Egede, son of the Greenland missionary Hans Egede, who in 1721 "re-discovered" Greenland. They originated from the West Greenland towns Nuuk and Qasigiannguit (Christianshaab). The first scientific collections were made by Morten Wormskjold in 1812-14 and Jens Vahl, who in 1830-36 collected in West Greenland. Together with later, more sporadic collections, not least the large Jens Vahl collections formed the basis of "Conspectus Florae Groenlandicae" (Lange 1880) in which is given a summary of all collections then available. Most important to the botanical exploration of West Greenland was the building in 1906 of the "Arctic Station" at Qeqertarsuaq (Godhavn) on the south coast of Disko, probably the floristically most diverse place in Greenland apart from the subarctic interior of some South Greenland fiords. Over 40 years the founder, M. P. Porsild and his sons, especially A. E. Porsild, collected numerous plants mainly in the Disko Bugt area. All collectors who had worked between Sisimiut and Nuussuaq were discussed by M. P. Porsild (1920), and Böcher (1963) brings the list up to date, extending it slightly southwards to Maniitsoq. In the first decades after World War II systematic botanical investigations were intensified from Disko southwards mainly by T. W. Böcher and co-workers, and in the Nuussuaq-Svartenhuk area mainly by K. Jakobsen who participated in four of the geological expeditions under the leadership of A. Rosenkrantz, who like other geologists brought home large collections, often from remote, high altitude localities. C. A. Jørgensen, T. Sørensen, and M. Westergaard, who collected along the west coast of Greenland in 1947, summarized the taxonomy and cytology of Greenland plants (Jørgensen & al. 1958).

Greenland Botanical Survey (G.B.S.) has been working in West Greenland northwards to the south coast of Nuussuaq for 22 summers. Dot maps based only on herbarium specimens often give a very misleading picture of the distribution of a taxon, especially common species that are only exceptionally collected. This was clearly illustrated by a map of the distribution of *Betula nana* in Greenland, prepared in 1980. Judging from this, the species clearly avoided the continental inland as only one, high altitude collection had been made within a 100 km broad zone along the ice cap from 65°45' to 68°N. Actually, *B. nana* is the characteristic plant of continental heaths, being with *Salix glauca* the most frequent species of all in this area. Because of this, all G.B.S. expeditions in the past two decades have collected all species of phanerogams within walking distance from base camps, which are usually in place from four to ten days. Consequently, as compared to earlier maps, those given in the present paper better reflect the actual distributions, yet as seen on Fig. 4 still some areas are insufficiently explored, esp. the inland from 65°-67°N, the two broad, east-west running areas near 67°30' and 68°15'N, and the archipelago north of 72°N.

4. Material and methods

Only specimens seen have been used in the mapping. More than 55,000 collections from West Greenland, kept in Herb. C, and a few additional collections from AAU, were examined. Of these, however, some were omitted. For quite many older collections the exact position of the locality is uncertain, e.g. many Vahl collections, labelled Distr. Colon. Umanak, and collections of the missionary J. F. D. Tietzen labelled Lichtenfels. Among the latter, many appear to have been collected not only at the missionary station (at 63°04'N) but also on his travels in the area (M. P. Porsild 1935). Likewise, the many collections of the Rev. P. H. Sørensen, who served in Illulissat but travelled in West and South Greenland in the last decades of the 19th century were omitted, as it is beyond doubt that many of his collections are mixed (M. P. Porsild 1920).

Recently introduced weeds and a few garden escapes in towns and settlements are not mapped here. However, maps of the total Greenland distribution of such taxa are given in Pedersen (1972), and South Greenland maps of some are included in Feilberg (1984).

No specimen list has been made, but the 54,784 sheets seen and used to prepare the maps are stamped, and for each locality the number of collections have been listed. Four maps (1:1,250,000) were used, and the dots were later transferred to the 1:2,500,000 map, reproduced in this paper in very reduced size. As the dotting was done by hand, no map showing number of taxa per locality was made, but the map of number of collections per locality (Fig. 4) gives an impression of the collecting intensity, which should still be kept in mind when studying the distribution maps.

A total of 963 localities have been recognised. On one third (319 loc.) less than 10 collections (not species) have been made, while on another third (339 loc.) 10-49 coll. were gathered. These are not included in Fig. 4, in which the smallest dots show the positions of 91 localities with 50-74 collections, and 52 localities with 75-99 collections. The next dot size marks 63 localities with 100-149 and 51 localities with 150-199 collections. The second largest dot marks 25 localities with 200-249 and 7 with 250-299 collections. At 16 sites more than 300 collections have been made, viz. 6 with 300-399 collections, 5 with 400-499, and 5 with even more, viz. Marrait on the western end of Nuussuaq (525 collections), Ikorfat on the north side of Nuussuaq (704), Sisimiut (827), Kangerlussuaq (1009), and Qeqertarsuaq (2500 collections). For comparison, a count based on the 379 maps gave for these five localities 127, 141, 191, 184 and 212 taxa resp. (Fig. 5). Marrait was the base camp for geological expeditions over several years, and Ikorfat another important geological base camp, from which T. Sørensen in 1947 collected plants of many species in connection with his fixations for chromosome studies. Sisimiut and Kangerlussuaq are important traffic centres, the latter being the only West Greenland international airport. The Arctic Station of the University of Copenhagen is located at Qeqertarsuaq.

5. Taxonomical considerations

Generally, the taxonomy follows Böcher & al. (1978). Exceptions are discussed.

Andromeda polifolia L.
In Greenland the species is represented by ssp. *polifolia*, found only on an island at 68°47'N in West Greenland, and by ssp. *glaucophylla* (Link) Hult., collected at four localities in Southwest Greenland between 60°56' and 63°22'N. All are included in the same map.

Antennaria Sect. Alpinae
Maps of *A. glabrata* (J. Vahl) Greene and *A. angustata* Greene have been made. The common distribution of the two taxa, incl. the almost exclusively alpine occurrences south of 67°N, and their frequent occurrences together, often with intermediate forms in snow-patches in the Disko Bugt area, may indicate that *A. glabrata* is a more or less glabrous extreme of *A. angustata*. Generally, the basal leaves of *A. angustata* are densely tomentose on the under side, whereas the hairiness of the upper side ranges from almost glabrous to tomentose. In *A. glabrata* the basal leaves are (nearly) glabrous on both sides, yet not even the type specimen of *A. glabrata* (in Herb. C) is totally glabrous. Its basal leaves have tiny ciliate hairs along the margin, and the stems have the same type of hairs. The stem leaves also have some webby hairs towards the base.

A. boecheriana A. E. Pors. This species, originally described as *A. canescens* (Lge.) Malte var. *pseudoporsildii* by Böcher (1963) but raised to a species by A. E. Porsild (1965), is included in *A. canescens*.

A. compacta Malte. The few, atypical specimens labelled *A. compacta* are included in *A. ekmaniana* A. E. Pors.

A. ekmaniana A. E. Pors. In Böcher & al. (1968, 1978) the achenes are said to be glabrous. This is not always the case, as also stated by A. E. Porsild (1965). None of the two South Greenland occurrences (Feilberg 1984, map 314) is typical *A. ekmaniana*, one being *A. canescens*, the other greatly deviating from this high arctic species, which otherwise has its southern limit at 67°N in alpine areas.

A. sornborgeri Fern. and *A. subcanescens* Ostf. are included in *A. canescens* (Lge.) Malte.

Antennaria Sect. Dioicae.
The closely related *A. hansii* Kern. and *A. intermedia* (Rosenv.) M. P. Pors. have been mapped separately, the main distinguishing character being pink-whitish versus the olive-brownish colour of the involucral bracts. However, in some collections the colour differs greatly even within the same inflorescense, and other distinguishing characters, e.g. number and length of stalks of capitulae and colour and densitiy of hairiness of basal leaves, are even less consistent. According to Böcher (1963) *A. hansii* is mainly an inland species of dry lichen heaths, while *A. intermedia* is a coastal species of herb-slopes and snow-patches. The large number of collections now at hand establish that they have an almost identical distribution in South Greenland (Feilberg 1984, maps 196 and 179) and in West Greenland. The few Southeast Greenland collections, all labelled *A. hansii*, show the same variation. Consequently, these two endemic Greenland taxa could as well be considered one species.

The third endemic species of Sect. Dioicae, *A. affinis* Fern., is usually easily distinguished. It has a pronounced continental distribution, reflecting its preference for dry, calcareous soil or loess. Scoggan (1979) considers the three taxa as ?microspecies of *A. rousseauii* Pors.

Armeria
Armeria is represented in South and Southwest Greenland by a species of salt-marshes (*A. maritima* (Mill.) Willd. ssp. *maritima*) and a circumgreenlandic species of fell-fields, heaths, and grasslands, rare and largely alpine in South Greenland (*A. scabra* Pall. ssp. *sibirica* (Turcz.) Hyl.). Iversen (1940) showed the first to be dimorphic. In some plants the pollen exine has a coarse reticulum, while the stigmas of the same plants have fine papillae, matching the fine pollen reticulum on other plants, which then have stigmas with coarse papillae. As the latter species is monomorphic, having a coarse reticulum on pollen, and papillae of same size, he considered them as two species, in contrast to Scoggan (1978) who treats them as varieties. None of the West Greenland plants was shown by Iversen to be dimorphic. However, a few, recent salt-marsh collections from Godthåbsfjord, Disko Bugt, and Nuussuaq resemble the South Greenland plants, yet their pollen is of the "*labradorica* " type (= *A. scabra* ssp. *sibirica*) (Iversen, 1940, Fig. 6). These are included in the map of *A. scabra*.

Betula
The total Greenland distribution of the three species of *Betula* and their hybrids have recently been mapped (Fredskild 1991). Eleven collections of *B. nana* L. x *B. glandulosa* Michx. are registered, all south of 66°N. They are not included in the maps here.

Calamagrostis hyperborea Lge.
According to Böcher & al. (1978) this South Greenland species is found south of 67°N. However, Feilberg (1984) revised and renamed the material, rejecting all finds north of 62°13'N (l.c., map 159: *C. inexpensa* A. Gray var. *robusta* (Vasey) Stebb.).

Calamagrostis poluninii Th. Sør.
This supposedly apomictic species, which was described by Sørensen (1954), is closely related to *C. purpurascens* R. Br. Collections best matching the description are mapped separately. One of the distinguishing characters is the shorter awn in *C. poluninii*, yet none of the characters mentioned is conclusive. In collections of true *C. purpurascens* there is a marked tendency for the awn to be shorter southwards. Contrary to Sørensen's statement the distribution area overlaps with that of *C. purpurascens*, except for a single collection of *C. poluninii* in South Greenland where *C. purpurascens* is missing.

Carex bigelowii Torr.
In South and Southwest Greenland this common, highly variable species occurs mainly as two, not too well dis-

tinguished ssp., viz. ssp. *nardeticola* Holub on heaths, and ssp. *bigelowii*, growing along streams and in other wet habitats. No attempt to separate these was made. No less than seven hybrids are mentioned in Böcher et al. (1978).

Where *C. stans* Drej. and *C. bigelowii* occur together north of 67½°N. the distinctions between the two species cause difficulty. Often the utriculi are empty even in late summer (?hybrids), but, if available, the size and shape of the nuts can be used (Fredskild 1978). In *C. bigelowii* the nuts are narrowly obovate-lanceolate, flat, mostly bright coloured, and 1.35-2.05 x 0.9-1.45 mm; in *C. stans* the nuts are mostly dark brown, smaller (1.2-1.5 mm), broadly obovate, truncate above, somewhat swollen, and with a more or less marked keel on one side, and thus resemble a nut of a tristigmate *Carex*. Without nuts, the lowest bract (slender, shorter than or almost equalling the inflorescence, versus robust, equalling or longer than the inflorescence), and the growth form (creeping rhizomes versus tuft-like) have been used. The difference in epidermis of the upper side of the leaves (smooth versus papillose) is not diagnostic within the study area. Quite many collections of supposed hybrid origin have been excluded, yet some may have been mapped, mainly as *C. bigelowii*. Especially towards the southern limit of *C. stans* care has been taken not to include dubious specimens.

Carex capillaris L. coll.
By far most Greenland material can be referred to ssp. *fuscidula* (Krecz.) Löve & Löve, which almost always has a staminate terminal spike and 2-3 mm long perigynia, that abruptly taper to a long, glabrous beak. In continental parts

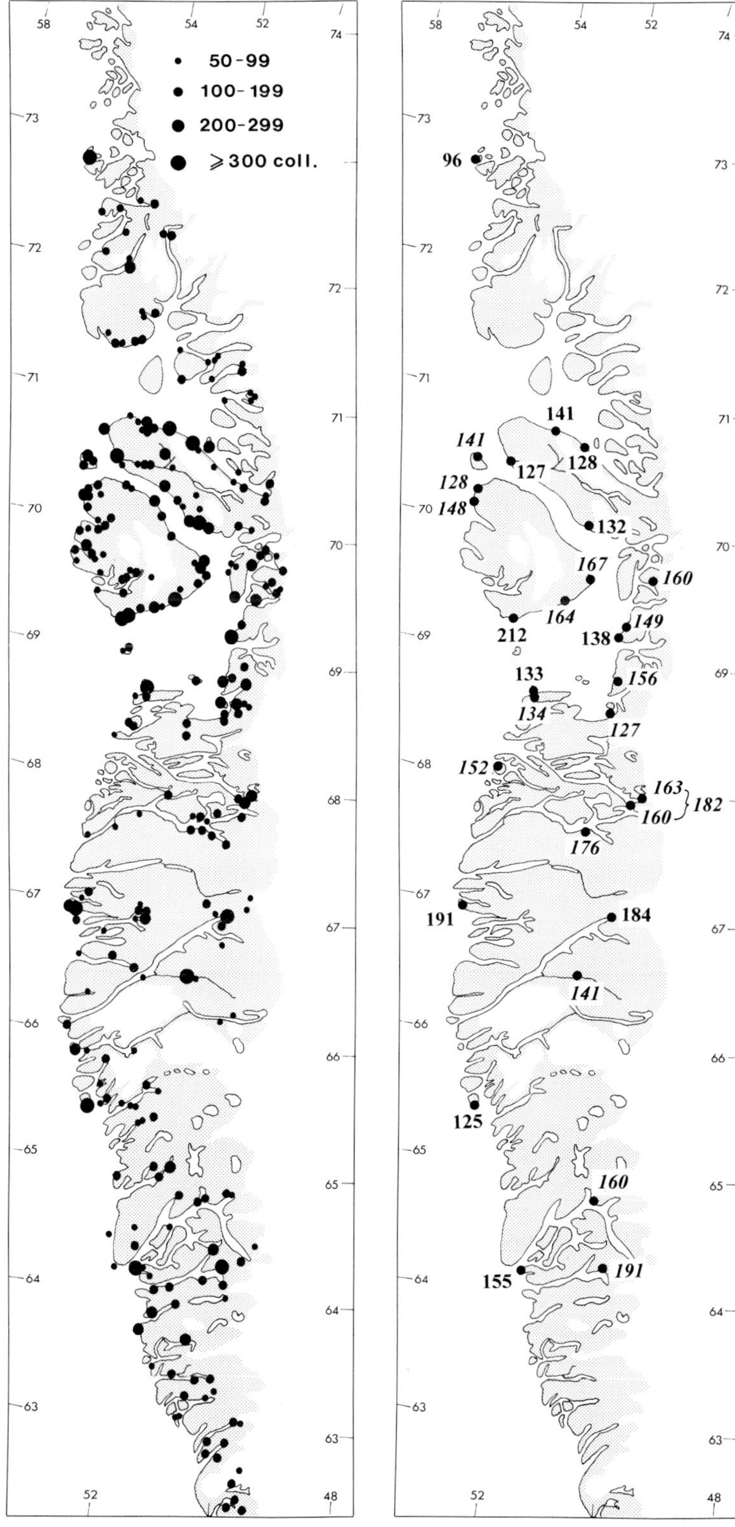

Fig. 4. Map showing the collecting intensity. Only localities with at least 50 collections (not species) are shown.

Fig. 5. Map of the number of taxa at some of the richest localities. Numbers in roman denote localities where collections have been made throughout time, whereas localities with numbers in italics denote localities where collecting were made during longer lasting Greenland Botanical Survey camps. Here, all taxa seen have been collected.

Fig. 4. Fig. 5.

of West and East Greenland is found a more robust form, ssp. *robustior* (Drej. ex Lange) Böch., with mostly gynaecandrous terminal spike and 2-3 mm long perigynia that gradually tapers to the beak, which often has small spines. The two ssp. have been mapped separately. Subspecies *porsildiana* (Polunin) Böch., with small (1.5-2.0 mm) perigynia, occurs in South Greenland. In Southwest Greenland many collections are intermediate between ssp. *fuscidula* and ssp. *robustior*, which speaks against treating the latter as a species: *C. boecheriana* Löve, Löve & Raymond. Such intermediates have been mapped as ssp. *fuscidula*. Surprisingly, Scoggan (1978) includes ssp. *robustior* (*C. boecheriana*) in ssp. *porsildiana*, which otherwise is characterized by small perigynia.

Carex capitata L.
The two Greenland taxa have been treated as species (Porsild & Cody 1979), subspecies (Böcher & al. 1978), varieties (Scoggan 1978), and forma (Raymond 1949). The main distinguishing characters dividing the West Greenland material have been: ssp. *arctogena* (H. Smith) Hiit., achenes mostly oblong ovoid, to 2.0 (-2.4) mm long (less beak), with many spines on the upper margin, and glumes as broad as the achene, or almost so; ssp. *capitata*, achenes broadest at the base, 2.2-2.9 mm, without spines or at the most only a few achenes in the spike having a few spines on the margin, glumes essentially smaller than the achenes. Four intermediate collections (67°46'-68°07'N) have not been mapped.

Carex marina Dew.
The map includes two closely related ssp. *marina* and ssp. *pseudolagopina* (Th. Sør.) Böch., discussed in Böcher (1952).

Cerastium alpinum L.
Böcher (1977) divides the *C. alpina*/*arcticum* complex into: *C. alpinum* L. ssp. *lanatum* (Lam.) Asch. & Graebn. and *C. arcticum* Lge. with three ssp.: *arcticum*, *procerum* (Lge.) Böch., and *hyperboreum* (Tolm.) Böch., whereas Scoggan (1978) subdivides *C. arcticum* only into varieties. Jørgensen & al. (1958) includes *C. arcticum* in *C. alpinum* L. coll.

Towards the south of Greenland *Cerastium alpinum* ssp. *lanatum* is most frequent in the lowland, whereas *C. arcticum* prefers alpine localities, in accordance with the separation in South Greenland (Feilberg 1984, maps 65 and 62). Bay (1992) treats all North Greenland material as *C. arcticum*, yet he mentions that some collections in the southern parts seem to be more related to *C. alpinum*. The large West Greenland material consists of c. 900 sheets, many of which can be matched to the different taxa. Maps for these have been made. However, too many indeterminable collections were left over, including some well-defined types, e.g. fairly glabrous plants growing in willow scrubs that resemble *C. arcticum* but have white, 5-10 celled hairs on the basal leaves. Consequently, only one map, including all collections of the complex, is given.

Draba alpina L. coll.
Böcher & al. (1978) and Bay (1992) divided the Greenland material of this complex into *D. alpina* L. and *D. bellii* Holm, whereas in Scoggan (1978) the two taxa are considered varieties of *D. alpina* L., viz. var. *nana* Hook. (including *D. bellii*) and var. *alpina*. The West Greenland material has been separated mainly on the shape of the pod (glabrous, lanceolate, acute in both ends in *D. alpina*; broader, often larger and always hairy in *D. bellii*). The leaves of *D. alpina* are almost glabrous, at least on the one side, with simple hairs on the edge, contrary to the more hairy leaves of *D. bellii*. The latter is a high arctic taxon. Towards its southern limit it grows mainly at higher altitudes. Thus, on Disko and Nuussuaq, 33 collections are from above 500 m, 9 from 200-500, and only 6 are from lower than this, mainly along rivers, which may have carried its seeds down. In contrast, *D. alpina* is a middle arctic taxon, with lowland as well as alpine occurrences throughout its distributional area.

Draba böcheri Gjærevoll & Ryvarden.
This species was described from material collected on the J. A. D. Jensen Nunatakker, which penetrates the Inland Ice (62°47-51'N, Gjærevoll & Ryvarden 1977). It is closely related to *D. lactea* Adams. Böcher & al. (1978) consider it a deviating type of this species.

Draba cinerea Adams and *D. arctica* J. Vahl.
Böcher (1966) discussed the Greenland material of this complex, separating *D. cinerea* Adams and *D. arctica* J. Vahl, with three subspecies, *groenlandica* (Ekm.) Böch., *arctica*, and *ostenfeldii* (Ekm.) Böch. In Böcher & al. (1978) ssp. *groenlandica* is given species rank: *D. groenlandica* Ekm. In contrast, Scoggan (1978) included all in *D. cinerea*. The West Greenland material falls into two fairly well-defined groups. In *D. cinerea* basal leaves are densely covered by tiny, greyish, stellate hairs, cauline leaves are 2-4, petals 2-3 mm, and seeds 0.6-0.8 mm. Typically, stellate hairs on basal leaves of *D. arctica* are slightly larger, and besides, the leaves are always ciliate along the lower edge. The cauline leaves are absent or one, petals 3-4 mm, and seeds 0.8-1.1 mm. Within *D. arctica* several forms are obvious, especially because of differences in form of pod and length of pedicel, but no attempt has been made to separate ssp. *groenlandica*, particularly as a collection (Godhavn, leg. Lagerkranz 1934, in Herb. C), shown in Böcher (1966, Plate V, d-f) as an example of *D. arctica* ssp. *groenlandica* is definitely *D. norvegica* Gunn. Some collections from Nuussuaq, Marmorilik, and Kangerdluarssuk (70°-71°N) are intermediate between *D. cinerea* and *D. arctica* and have been mapped as *D. arctica*. Plantae Vaculares Groenlandicae Exsiccatae No.

537 (Disko, Nordfjord, Aug. 10, 1975) has been distributed as *D. cinerea*. The sheet in Herb. C. is a *D. arctica*.

Draba norvegica Gunn. and *D. arctogena* Ekm.
About this complex Böcher (1966) writes, "In many respects the transition from *D. norvegica* to *D. arctogena* is clinal". The West Greenland material is extremely variable, especially in hairiness, but also in the number of cauline, more or less dentate leaves, and the shape of the pod, which in the southern part often is very long and narrow, but broader and more or less ovate in plants from farther north. Consequently, all material is considered *D. norvegica* Gunn.

Draba lactea Adams and *D. fladnizensis* Wulf.
The few West Greenland collections of *D. fladnizensis* Wulf. have been separated from *D. lactea* Adams on the following criteria: leaves with a few simple or (rarely) only once forked hairs, almost exclusively marginal; stems often many, stiff, reddish, totally glabrous, usually with one leaf; flowers several in a racemose (not corymbose) inflorescence; and pods long, narrow with almost parallel sides.

Many collections of hybridogenous character, presumably with *D. lactea* as one of the parents, were found. They are not included in the map.

Draba subcapitata Simm.
Contrary to the North and Northeast Greenland material, West Greenland collections of this species often have single, simple marginal hairs, sometimes also on the valves. Of the 38 collections with the altitude given 32 are from above 400 m (9 of these from 1000 m or above).

Erigeron uniflorus L. coll.
On the basis of colour and type of involucral pubescense and the shape of the capitulum base three species have been separated: *E. humilis* Grah., *E. eriocephalus* J. Vahl, and *E. uniflorus* L. The last possibly includes some *E. borealis* (Vierh.) Simm.

Eriophorum angustifolium Honck. ssp. *subarcticum* (V. Vassil.) Hult. and *E. triste* (Th. Fr.) Hadac & Löve.
These have been treated as varieties (Scoggan 1978) and species (e.g. Böcher & al. 1978). Main distinguishing characters are: peduncels long and glabrous; spathes and scales dull brown to lead-coloured, and leaves broad in *E. angustifolium* versus short and scabrous, black or blackish, and narrow in *E. triste*. Mainly on northern Disko, Nuussuaq, and Svartenhuk, but even as far south as Nordre Strømfjord (c. 67°30'N) intermediates otherwise resembling *E. angustifolium*, but with single, stiff hairs on the peduncels are found. These are mapped as *E. angustifolium*. A few collections were omitted, as they were in all respects intermediate.

Hieracium
The material has been mapped as *H. hyparcticum* Almq. (Sect. Nigrescentia, including *H. lividorubens* Almq.), *H. groenlandicum* (A.-T.) Almq. (Sect. Prenanthoidea, incl. *H. ivigtutense* (Almq.) Om.), *H. rigorosum* (Læst.) Almq. (Sect. Tridentata, incl. *H. acranthophorum* Om.), and *H. alpinum* (L.) Backh.

Hierochloë
In the key in Böcher & al. (1978) it is erroneously stated that the awns of the florets of *H. orthantha* Th. Sør. are scarcely exerted from the spikelet. Sørensen (1954) mentioned that *H. alpina* (Willd.) R. & S. and *H. orthantha* occur together in West Greenland across only one degree of latitude. The much larger material now at hand shows the area to be more than four degrees.

x *Ledodendron vanhoeffeni* Dalgaard & Fredskild
The hybridogenous character of x *L. vanhoeffeni* (*Rhododendron vanhoeffeni* Abromeit) has recently been discussed in Dalgaard & Fredskild (1993).

Lychnis alpina L. ssp. *americana* (Fern.) Feilberg
Feilberg (1984) treated *Viscaria alpina* (L.) Don var. *americana* as a subspecies.

Phippsia algida (Sol.) R. Br.
Besides the widespread *P. algida*, Bay (1992) recently showed that ssp. *algidiformis* (H. Sm.) Löve & Löve occurs in northern Greenland. In West Greenland it is scattered between south Disko and Upernavik, the last mentioned locality only represented by a Vahl collection "Prope Colon. Upernavik". Plants from the Disko-Nuussuaq area may have some hairs and violet nerves (othervise characterizing ssp. *algidiformis*) on some or most lemmae which, however, are only 1.2-1.6 mm long (versus 1.5-2.3 mm long in ssp. *algidiformis*), thus indicating *P. algida*. These plants also have the rather lax growth form of P. *algida* in which they are included.

Poa abbreviata R. Br.
Of the 17 collections of this species 14 are typical, three slightly deviating. In West Greenland they have the same, very restricted distribution as *P. hartzii*.

Poa arctica R. Br. and *P. pratensis* L.
Typical plants of *P. arctica* R. Br., with large, open panicles with slender branches, and usually 1(-2) spikelet(s), are common in N.Greenland. Southward in West Greenland such plants are found mainly inland and – in South Greenland – mainly at high altitude. On the contrary, typical *P. pratensis* L. ssp. *alpigena* (Blytt) Hiit. (panicle contracted, several spikelets on branches) is very common in South Greenland, fairly frequent in West Greenland to c. 74°N, and with scattered occurrences northwards to c. 78°N (Bay 1992, map 120). However, West Greenland material, exceeding 1100 col-

lections labelled as one of these two taxa, is extremely variable, with numerous intermediates, as illustrated by a collection from Tasiusak (73°25'N), leg. Thorild Wulff, 1916, and determined by C. H. Ostenfeld. It was later redetermined/renamed by Johs. Lid 1931, P. Gelting 1933, Th. Sørensen 1936, J. A. Nannfeldt 1937, and C. Bay 1982. Consequently, only one map including all material is given.

The distribution of the viviparous *P. pratensis* var. *colpodea* (Fries) Schol., found almost exclusively at high elevations on Nuussuaq and Svartenhuk, is shown on a separate map.

Poa hartzii Gand.
Thirty-one collections, all from Disko and the western half of Nuussuaq, have been determined as *P. hartzii*, while 22 collections from the same area and, with one exception, from the same localities, are less typical and therefore are omitted. Some of these resemble *P. glauca* M. Vahl.

Poa nemoralis L.
The main distinguishing characters used to separate this species from the highly variable *P. glauca* M. Vahl are the length of ligule (below 0.5 versus 1.2-2 mm) and size of spikelet (less than 4 versus 4-6 mm).

Poa flexuosa Sm.
This species was first found at c. 1000 m altitude on Akuliaruserssuaq in South Greenland by Polunin (1943). It was rejected by Jørgensen & al. (1958) but found again on the J. A. D. Jensen Nunatakker in Southwest Greenland (Gjærevoll & Ryvarden 1977) and confirmed by J.A. Nannfeldt. It is not mentioned in Böcher & al. (1978) and Feilberg (1984). In 1989 S. Lægaard examined the Greenland material of *Poa glauca* in Herb. C and found quite many of *P. flexuosa.* Most of these plus some collections from Herb. AAU and some recent coll. in Herb. C are included on the map. The known distribution ranges from Ingitait Fjord (61°09'N) in Southeast Greenland to Ikorfat (70°45'N) in West Greenland. With the exception of one of the three collections from the northernmost locality, all collections with the altitude given are alpine (450-1400 m).

Potentilla nivea L. s.l.
In Greenland *P. nivea* L. emend. Hult. is a southern taxon, contrary to *P. hookeriana* Lehm. s.l. (Bay 1992, maps 95 and 50). Almost all petioles in specimens from the southern part of West Greenland are floccose with curly hairs. From c. 64°N a few, stiff, shiny or papillouse hairs occur occasionnally, and from c. 66°N such hairs are more frequent, but as far north as 72°N plants with typical "*nivea*-petioles" occur. Plants with floccose hairs all dominating are considered *P. nivea* and only these are mapped.

Within *P. hookeriana* aut. some collections have largely more or less stiff, long hairs on the petiole (ssp. *chamissonis* Hult.), others also a dense cover of more or less stiff, short hairs (ssp. *hookeriana*). However, by far most collections are intermediate and almost always with some floccose hairs. Two tentative maps of the most typical collections of ssp. *chamissonis* and ssp. *hookeriana*, respectively, showed completely congruent distributions. Consequently, only one map is presented: *P. hookeriana* L. s.l. It is almost identical with a tentative map of intermediates between *P. nivea* and *P. hookeriana.*

Puccinellia
The Greenland *Puccinellia* have been thoroughly studied by Sørensen (1953) who also, to the best of my knowledge, developed the key presented in the first two editions of "Grønlands Flora", incl. the English version (Böcher & al. 1968). Some species can safely be determined, whereas others are most difficult, and much material in Herb. C remains undetermined. The troubles are illustrated by some measurements of the two xerophytic species, *P. deschampsioides* Th. Sør. and *P. groenlandica* Th. Sør., between which intermediates occur, as already stated by Sørensen (l.c., p. 32). The length of the first glume given in the descriptions in Sørensen (l.c.) is 1.1-1.5 mm in *P. deschampsioides* versus 2.0-2.5 mm in *P. groenlandica*, of the second glume 1.8-2.1 versus 2.5-3.0 mm, of the anthers 0.7-0.9 versus 1.1-1.5 mm. In many collections, including some determined by Sørensen to one or the other species, the measurements are 1.6-1.8 mm for the first glume, 2.3-2.4 for the second, and 0.8-1.1 mm for anthers.

As a consequence of the many difficulties, K. Jakobsen revised the key in the third edition of the flora (Böcher & al. 1978). *Puccinellia groenlandica* is found twice, under "keel of palea with long hairs towards the base, with small spines above", and under "keel of palea glabrous or almost glabrous towards the base, with small spines above", yet the measurements of glumes and anthers quoted above are retained.

Likewise, the separation of *P. coarctata* and *P. vaginata* may cause great difficulties, illustrated by Sørensen's description of an intermediate form, *P. vaginata* var. *paradoxa* Sørensen, which is "connected with P. vaginata by an unbroken series of integrading forms" (Sørensen l.c.:47).

In the maps only reasonably typical specimens have been noted.

Puccinellia laurentiana Fern. et Weath.
According to Sørensen (l.c.:41) only one deviating collection from Greenland is at hand. It was collected by Lagerkrantz in 1936 at Eqaluit ilordlit (64°09'N) at the head of Ameralik Fjord, not at Eqaluit (64°03'N) at the mouth of the fiord as stated by Sørensen. The species is disregarded here.

Puccinellia porsildii Th. Sør.
This species is only known from the type locality, an abandoned chicken-run at the Arctic Station on Disko. The type was collected in 1933 by M.P. Porsild, more specimens in 1947 by T. Sørensen. The species is disregarded here.

Puccinellia rosenkrantzii Th. Sør.
This species seems closely related to *P. deschampsioides*. It grows at a few sites on Nuussuaq (Sørensen l.c. map p. 174), on fresh deposits around active mud volcanoes. More distant from these grows *P. deschampsioides* with which it forms intermediates (l.c.:35), as confirmed by the collector of the type specimens and other collections (K. Jakobsen, pers. comm.).

Sagina intermedia Fenzl and *S. caespitosa* (J. Vahl) Lge.
Most often plants of these two species are easily determinable, but intermediates may be difficult to separate, as indicated by Scoggan (1978), who only treats them as varieties (*S. nivalis* (Lindbl.) Fries var. *nivalis*, and var. *caespitosa* (Vahl) Boivin). Distinguishing characters (Böcher & al. (1978), Scoggan (l.c.), Porsild & Cody (1979), and Tutin & al. (1964)) are: sepals 4(-5), 1.5-2.0 mm long in *S. intermedia* versus sepals 5, 1.8-3 mm long in *S. caespitosa*; petals 4 (mostly), shorter than sepals versus petals 5, longer than sepals; stamens 4-5 (Tutin et & al.: usually less than 10) versus 10. In the West Greenland material the following distinguishing characters have been used: *S. intermedia* has typically a distinct, central rosette of slightly broader leaves and long pedicels carrying small flowers with 4 petals, rounded sepals, and 7-8 stamens. Old specimens may form a cushion but the root is branched right at the top. Plants of *S. caespitosa* are typically in cushions with the tap-root not branched at the top; pedicels are short, flowers larger, usually have 5 sepals and petals and 10 stamens (or almost). Smaller specimens resembling *S. intermedia* are not uncommon. These have been disregarded in the mapping. Four- and 5-merous flowers occur on the same plant in some collections, and often the length of petals equals that of sepals, varying from shorter to longer on the same plant.

Salix
Contrary to *S. arctica* Pall., branches of *S. glauca* L. coll. never root. Towards the southern limit of the former only collections combining dark, two-coloured catkin-scales with rooting branches have been mapped as *S. arctica*.

Saxifraga rivularis L. s.l.
In the treatment of North Greenland (Bay 1992) and South Greenland plants (Feilberg 1984) *S. hyperborea* R. Br. has been mapped separately. The West Greenland material can readily be separated in two groups, viz. typical *S. rivularis* with rooting stolons, mostly growing on mossy ground at streams, seashores, birdcliffs, and inhabited places, and typical, reddish coloured *S. hyperborea* without stolons, mostly from high elevations or N-slopes. However, a third pile of intermediate (or poor) collections is even taller, and consequently only one map is prepared, in accordance with Scoggan (1978) who treats *S. hyperborea* as a forma.

Saxifraga nivalis L. and *S. tenuis* (Wbg.) H. Sm.
There are sometimes problems in the separation of these two species. However, in most cases the separation seems safe, which perhaps is a consequence of the different chromosome numbers (2n = 60 and 20, resp.). Therefore, two maps have been prepared. Most of the intermediate collections and their identification is based on general impression only. Only a few collections remain undetermined.

The separating characters are: stems of *S. nivalis* usually green, fairly thick, with white (or slightly reddish), curly hairs even to the base, but numerous on the upper part versus stems in *S. tenuis* usually reddish, slender, hairs (almost) always reddish, stiff, and shorter, very few or none toward the base; inflorescences in *S. nivalis* (especially towards the north) a terminal head of sessile flowers, large individuals, and towards the south also smaller individuals, often cymose, but usually with more flowers on each peduncle versus in *S. tenuis* open, few-flowered with only one flower on each, fairly long peduncle; ripe carpels in *S. nivalis* with the tip bent outwards or slightly recurved versus in *S. tenuis* always very recurved.

Common to most species of *Saxifraga* as well as to species of other genera is the tendency to become more reddish at higher latitudes. Thus, many *S. tenuis* in South Greenland have green stems with bright hairs.

Scoggan (1978) and Porsild & Cody (1978) mention as a distinguishing characteristic the absence / presence of coarse, rust-coloured hairs on the underside of the leaves and on the petioles. This is not valid for the West Greenland material.

Stellaria longipes Goldie s.l.
According to Böcher (1951), Philipp (1972), and Böcher & al. (1978) this aggregate species is represented in Greenland by *S. longipes* Goldie s.str., *S. edwardsii* R. Br., *S. crassipes* Hult., *S. monantha* Hult., *S. laeta* Richards, and possibly *S. laxmannii* Fisch. As discussed by Bay (1992), typical specimens of these taxa have different distributions, but as so many intermediates occur mapping the major part of the material is impossible. Consequently, only two maps have been prepared: *S. longipes* Goldie s.l. "A", sepals glabrous (incl. *S. longipes* s.str., *S. monantha*, and *S. crassipes*), and *S. longipes* Goldie s.l. "B", sepal margins ciliate (incl. *S. laeta*, *S. edwardsii*, and possibly *S. laxmannii*). The latter map includes collections in which the sepal margins are ciliate and more or less hairy on the back.

Taraxacum
Three taxa have been separated: *T. lacerum* Greene (incl. *T. umbrinum* Dahlst.) with a conspicuous corniculate appendage near the tip of the phyllaries, of which the outer ones are appressed, *T. croceum* Dahlst. (incl. *T. amphiphron* Böch.) with recurving outer phyllaries and without (or almost) appendages, and *T. phymatocarpum* Vahl.

Trisetum
T. triflorum (Bigel.) Löve & Löve (incl. ssp. *triflorum* and ssp. *molle* (Hult.) Löve & Löve) has been separated from *T. spicatum* (L.) Richt. based on its loose, greenish panicle and the length of the stamens. Only typical specimens are mapped as *T. triflorum*. The awn, which is straight early in summer in contrast to that of *T. spicatum*, tends to become geniculate later. Ecologically the two taxa differ, *T. triflorum* being a southern species preferring south facing-slopes, with e.g. *Carex supina*, and, towards the north, with open *Salix* scrubs. On the contrary, *T. spicatum* is a northern species of snow-patch communities, and towards the south preferring high altitudes.

Veronica
The separation of *V. alpina* L. and *V. wormskjoldii* R.& S. is based on the hairiness of the capsules and calyx lobes. *Veronica alpina* has glabrous capsules and the calyx lobes are only hairy in the margin, whereas *V. wormskjoldii* has hairy capsules and hairs on the back of the calyx lobes. Capsules of three collections had only single hairs on the upper end of the capsule. They are included in *V. alpina*, as are possible collections of its var. *australis* Wbg.

6. Results and discussion

6.1. Distribution types

The distribution maps of the 379 taxa have been grouped in 11 West Greenland distribution types (WGDT), most of which are subdivided. For convenience the western half of Greenland is divided into: West Greenland, defined as the investigation area (62°20' – 74°N); South Greenland, the area south of 62°20' N; Northwest Greenland, the area from 74°N to Humboldt Gletscher at c. 79°N; and North Greenland, north of this (cfr. Bay 1992, Fig. 23). The taxa are arranged alphabetically within the types and subtypes. The labels of most old collections do not specify habitat and altitude, and such collections were disregarded when the altitudinal range of a taxon is mentioned.

In the upper, right corner of each map the total Greenland distribution is outlined, based on Feilberg (1984), Bay (1992), and Herb. C. Reference to publications, if any, which includes maps of total Greenland distribution, is given under each species in Böcher & al. (1968), and in the 1978-edition the reference list is brought up to date. Since then maps of the following species have been published: *Leymus* (*Elymus*) *arenarius*, *L. mollis*, and their hybrids (Ahokas & Fredskild 1991), *Festuca vivipara* ssp. *vivipara*, ssp. *hirsuta*, and ssp. *glabra* (Frederiksen 1981), *Festuca saximontana* (Frederiksen 1982), all species of *Epilobium* and *Chamaenerion* (Fredskild 1984b), *Betula*, incl. hybrids (Fredskild 1991), and the limnophytes: *Alopecurus aequalis*, *Eleocharis acicularis*, *Hippuris vulgaris*, *Juncus subtilis*, *Limosella aquatica*, *Menyanthes trifoliata*, *Pleuropogon sabinei*, *Ranunculus confervoides*, *R. reptans*, *Subularia aquatica*, and all species of *Callitriche*, *Isoëtes*, *Myriophyllum*, *Potamogeton*, and *Utricularia* (Fredskild 1992).

Type 1. Taxa occurring all over W.Greenland, having no limit in the area.

Subtype 1a (Maps 1-54). Taxa evenly distributed all over West Greenland. The maps are arranged in truly circumgreenlandic (1-17), circumgreenlandic less Melville Bugt in Northwest Greenland (18-20), almost circumgreenlandic yet missing in part of North Greenland or other parts (21-34), and southern taxa, absent from North Greenland and also often in the northern part of Northwest Greenland (35-54). Five taxa have their northern limit just north of West Greenland at 74°10-20'N, viz. *Betula nana*, *Campanula gieseckiana*, *Carex scirpoidea*, *Saxifraga paniculata*, and *Tofieldia pusilla* (Maps 35, 36, 39, 52, 53).

Subtype 1b (Maps 55-67). Taxa becoming rare towards south. Most of these taxa are very rare in South Greenland and are found mainly or only inland, often at high altitudes. *Phippsia algida* (64) mostly grows at the outer coast in South Greenland. With the exception of *Draba lactea* (60) and *Erigeron compositus* (62) taxa of this subtype are absent from or only rarely occur in the interior of North Greenland. *Antennaria ekmaniana* (55) only has one, alpine occurrence in South Greenland (61°54'N).

Subtype 1c (Maps 68-71). Southern taxa, missing in Melville Bugt and very rare in the interior North Greenland.

Subtype 1d (Maps 72-73). Taxa having their main Greenland distributional area within West Greenland.

Subtype 1e (Maps 74-85). Circumgreenlandic or southern taxa, otherwise widespread, yet missing or extremely rare in the highly continental inland around 67°N. Besides the two sea-shore plants, *Carex glareosa* (75) and *Puccinellia phryganodes* (82), and *Stellaria humifusa* (85), by far most frequent on sea-shores, this subtype includes taxa from herb-slopes and heaths on acid soil. Three taxa have their northern limit between 74°07' and 74°20'N: *Harrimanella hypnoides* (77), *Loiseleuria procumbens* (79), and *Phyllodoce coerulea* (81).

Subtype 1f (Maps 86-87). Two taxa avoiding the basaltic area 69°-72°N, one of which, *Diapensia lapponica* (86), also avoids inland neutral-alkaline soils at 67°N.

Type 2. High arctic taxa with southern limit in the Disko-Nuussuaq area.

Subtype 2a (Maps 88-99). In West Greenland only distributed in the basaltic area on north Disko-western half of Nuussuaq-Svartenhuk, some taxa also in the marble area at Marmorilik on the mainland. Some taxa occur in high altitudes only: *Draba adamsii* (90), and *Minuartia rossii* (94), or exceptionally also on steep screes or in a river delta: *Draba bellii* (91), *D. subcapitata* (92), *Poa pratensis* var. *colpodea* (97). Other taxa are found only in the lowland: *Braya purpurascens* (88), *B. thorild-wulffii* (89), *Poa hartzii* (96), and one is a seashore plant: *Puccinellia andersonii* (99).

Subtype 2b (Maps 100-108). Resembling 2a yet not restricted to basalt or marble. Some taxa also occur at one locality, *Ranunculus sulphureus* (107), *Taraxacum phymatocarpum* (108); or two, *Festuca hyperborea* (104) in Melville Bugt.

Type 3. Northern, i.e. high arctic and middle arctic, taxa with their southern limit between Disko Bugt (c. 68°30'N) and Maniitsoq ice cap (66°N).

Subtype 3a (Maps 109-119). Main West Greenland distribution in the Disko (Bugt)-Nuussuaq-Svartenhuk area. No occurrence in the inland at 67°N. Some are alpine towards their southern limit, *Erigeron eriocephalus* (112), *Potentilla hyparctica* (114), and *P. vahliana* (115). *Alopecurus alpinus* (109) is clearly anthropochorous towards its southern limit, growing on manured ground in settlements and on former camp sites.

Subtype 3b (Maps 120-124). Like 3a yet also occurring inland at 67°N. One, *Carex ursina* (121) is a seashore plant.

Subtype 3c (Maps 125-129). Towards the southern limit growing only inland, both in the lowland and at higher altitudes. *Antennaria ekmaniana* (55), mostly occurring at higher altitudes at 67°N, is closely related to this subtype, but because of one isolated occurrence in S.Greenland it is grouped in type 1b.

Subtype 3d (Maps 130-132). Mainly in the inland. *Dryopteris fragrans* (130) and possibly also *Tofieldia coccinea* (132) avoids basalt.

Type 4. Northern taxa with their southern limit between Maniitsoq ice cap and 62°20'N.

Subtype 4a (Maps 133-136). No preference as to altitude or degree of continentality. *Ledum palustre* ssp. *decumbens* (135) has its total Greenland distribution in the western half of Greenland.

Subtype 4b (Maps 137-147). Towards the southern limit only occurring inland. Two taxa almost only grow at high altitudes, *Arnica angustifolia* (137), *Pedicularis hirsuta* (145). One, *Cassiope tetragona* (140), is mainly found on north-facing slopes, often at high altitudes. Strictly speaking, the middle arctic *Erigeron humilis* (142) should be grouped in Type 6. However, its main distribution area is 67°-73°N and its two southernmost occurrences are high alpine at 62°50' and 61°N. *Halimolobus mollis* (143) has its total Greenland distribution in western Greenland.

Type 5. Three southern taxa (Maps 148-150) with their northern limit between Upernavik Isstrøm (c. 73°N) and 74°N. Phytogeographically they are similar to the eight taxa of Subtype 1a and 1e in having their northern limit at 74°07-20'N.

Type 6. Southern taxa with northern limit at Upernavik Isstrøm (c. 73°N).

Subtype 6a (Maps 151-164). Widely distributed taxa without preferences for bedrock type or degree of continentality. Two have their southern limit just south of West Greenland, viz. *Artemisia borealis* (151) at 62°14'N, and *Pedicularis lapponica* (160) at 62°05'N, and are thus similar to Type 11. One is a limnophyte (*Potamogeton pusillus*, 162).

Subtype 6b (Maps 165-167). Taxa mainly occurring inland in the southern part of West Greenland. In South Greenland they are found exclusively in the inland, *Draba crassifolia* (166) only at three alpine sites.

Subtype 6c (Maps 168-176). Taxa not found in the inland at 67°N. *Carex subspathacea* (170) is a salt-marsh plant, the others grow mainly or preferably in herb-slopes. *Juncus trifidus* (174) only exceptionally grows on basalt.

Subtype 6d (Maps 177-178). Taxa avoiding basalt. Both taxa have been found only once in the inland at 67°N.

Type 7. Southern taxa with northern limit at Svartenhuk (71°-72°N).

Subtype 7a (Maps 179-191). Taxa without preference to bedrock type and degree of continentality. *Potentilla egedii* (187) is limited to sea-shores.

Subtype 7b (192-195). Taxa avoiding the inland at 67°N. One is a herb-slope plant (*Veronica alpina*, 185).

Subtype 7c (196-200). Taxa mainly found in the inland. Two are limnophytes.

Type 8. Southern taxa with northern limit in the Disko-Nuussuaq area (69°-71°N).

Subtype 8a (Maps 201-217). Taxa without preference to bedrock type and degree of continentality, yet *Juniperus communis* ssp. *alpina* (210) avoids the outer coast of the mainland towards the north, as do *Angelica archangelica* ssp. *norvegica* (202) and *Epilobium palustre* (206). Without exception *Angelica* grows only at homothermic springs on Disko. Two localities on the mainland halfway between the northernmost dots and the outer coast between 67°N and 68°N carry the eskimo name "Kuanit" meaning the place where kvan (= *Angelica*) grows. The three westernmost occurrences of *Epilobium palustre* on Disko are at homothermic springs. No information on habitat is given with the eastern specimen. The isolated occurrence of *Draba incana* (205) on Disko is at a homothermic spring. With one exception all occurrences of *Poa flexuosa* (212) are alpine.

Subtype 8b (Map 218). Taxon avoiding basalt.

Subtype 8c (Maps 219-231). Taxa avoiding the inland at 67°N. Many grow in herb-slopes and willow scrubs, one is a limnophyte (*Callitriche anceps*, 221).

Subtype 8d (Maps 232-252). Taxa preferring the outer coast, especially towards the north on the mainland, where most are missing north of 67°N. Most are herb-slope or willow scrub plants. *Epilobium hornemannii* (238) and *E. lactiflorum* (239) almost exclusively grow in mossy vegetation at homothermic springs on Disko.

Subtype 8e (253-262). Taxa preferring the inland. Five are limnophytes.

Subtype 8f (Maps 263-266). Taxa with disjunct Greenland distribution. The isolated occurrence of *Hieracium alpinum* at Kuanit, Diskofjord (leg. Gelting 1949) has recently been searched for in vain (J. Feilberg, pers. comm.). Because of this it was disregarded in Feilberg (1984, map 221).

Type 9. Taxa with northern limit between Maniitsoq ice cap (66°N) and the south border of Disko Bugt (68°N).

Subtype 9a (Maps 267-270). Taxa occurring in coastal as well as inland areas. One is a limnophyte (*Callitriche hamulata*, 268).

Subtype 9b (Maps 271-272). Like 9a, yet not in the inland at 67°N. One is a limnophyte (*Myriophyllum alterniflorum*, 271).

Subtype 9c (Maps 273-286). Inland taxa, especially towards the north. Some are rare, with a disjunct Greenland distribution. Nine taxa have their total Greenland distribution in West and South Greenland. Three are limnophytes or restricted to lake-shores.

Subtype 9d (Maps 287-303). Coastal taxa, especially towards the north. Several are herb-slope plants.

Type 10. Taxa with northern limit at or south of Sukkertoppen Iskappe (66°N). 20 of the species have their total Greenland distribution in West and South Greenland.

Subtype 10a (Maps 304-324). Taxa occurring both at the coast and inland or without clear preference. One is a limnophyte (*Isoëtes echinospora*, 312), another a sea-shore plant (*Puccinellia maritima*, 318).

Subtype 10b (Maps 325-335). Taxa preferring the inland. One (*Carex salina*, 328) is restricted to, and one (*Ligusticum scoticum*, 332) mainly found on sea-shores.

Subtype 10c (Maps 336-341). Taxa preferring the outer coast areas.

Type 11. Taxa with their total Greenland distribution in West Greenland (27 taxa) or with their distribution in the western half of Greenland within West Greenland, but also occurring in East Greenland (11 taxa: 6 in subtype 11b, 1 in 11c, and 4 in 11d). *Spergularia canadensis* (374), *Cerastium arvense* (378), and *Pedicularis groenlandica* (379) are known from only one locality in Greenland.

Subtype 11a (Maps 342-346). Taxa occurring over several degrees of latitude, without preference to degree of continentality. One is a limnophyte (*Callitriche hermaphroditica*, 343). *Artemisia borealis* (151), only occurring in western Greenland but with southern limit at 62°14'N, and *Pedicularis lapponica* (160) with southern limit at 62°05'N but also occurring in East Greenland, are similar to this subtype.

Subtype 11b (Maps 347-358). Like 11a, yet preferring the inland, mostly occurring on rich soils. One is a limnophyte (*Utricularia ochroleuca*, 358).

Subtype 11c (Maps 359-362). Like 11a, yet preferring coastal areas; only one growing at sea-shores (*Puccinellia langeana*, 362).

Subtype 11d (Maps 363-367). Taxa occurring over one to a few degrees of latitude. *Epilobium arcticum* (366) avoids gneissic areas. *Puccinellia rosenkrantzii* (367) is endemic to Greenland.

Subtype 11e (Maps 368-375). Like 11d, yet avoiding the outer coast. x *Ledodendron vanhoeffenii* (371) is endemic to Greenland. One is a sea-shore plant (*Atriplex longipes*, 369), another the only marine vascular plant in Greenland (*Zostera marina*, 375).

Subtype 11f (Maps 376-379). Like 11d, yet preferring coastal areas.

In Böcher & al. (1959) most Greenland taxa were referred to one of ten Greenland climatic distribution types which Böcher (1963) extented to 11, now termed biological distribution types (BDT). In Böcher (1975) the list was revised and almost completed. Based on this, the percentage composition of the BDT for each of the 11 West Greenland geographical distribution types (WGDT) recognized here, has been calculated (Table 2). Clear correlations are seen, e.g., in Type 2 where only high arctic, arctic continental and middle arctic species occur, whereas in Type 5 only low arctic and low arctic oceanic species are found. Generally, in the West Greenland phanerogam flora one third of the species is high and middle arctic, one third is low arctic, and one third is boreal. For comparison, the corresponding percentages for North and South Greenland are given, illustrating the intermediate position of West Greenland (Bay 1992, Feilberg 1984).

In Table 3 the distribution types are compared with the geographical distribution types (GDT) as suggested by Hultén (1958, 1964, 1971). When compared with North and South Greenland the main difference is the larger number of circumpolar taxa in North Greenland, and of eastern taxa in South Greenland. Bay (1992) mapped 218 taxa for North Greenland. For South Greenland Feilberg

Table 2. Percentage composition of the biological distribution types according to Böcher (1975) in the West Greenland distribution types (WGDT, left column). A: Arctic, widespread; HA: High arctic; AC: Arctic, continental; MA: Middle arctic; L: Low arctic; LO: Low arctic, oceanic; LC: Low arctic, continental; B: Boreal; BO: Boreal, oceanic; BC: Boreal, continental; BS: Boreal, sylvicolous. Number of taxa in each type, see Table 3. Percentages from North and South Greenland are from Bay (1992) and Feilberg (1984), respectively.

WGDT	A	HA	AC	MA	L	LO	LC	B	BO	BC	BS
1	38	1	12	5	24	8	5	8			
2		90	5	5							
3	4	29	33	25			8				
4		7	57	21			14				
5					33	67					
6				7	29	43	11	4	4		4
7					23	14	23	41			
8					9	26	5	35	3	3	20
9					3	27	5	46	8	8	3
10					3	11		39	21	8	18
11			8	24	8		29	18	3	11	
n taxa	33	28	30	25	45	55	32	79	15	12	22
W.Grl. %	9	7	8	7	12	15	9	21	4	3	6
n taxa			116			132			128		
W.Grl. %			31			35			34		
N.Grl. %	17	21	17	11	13	9	7	5		0.5	
			66			29			5		
S.Grl. %			16			35			49		

(1984) mapped 346, of which several were recently introduced. As the species delimitation differs slightly between the present paper and the two mentioned, these numbers have not been included in the table.

In the list of taxa (Table 4) besides the map number, the West Greenland distribution type (WGDT), the biological distribution type (BDT), and the geographical distribution type (GDT) are given for each taxon. ( ) means uncertainty as to type; such taxa are not included in Table 2 and 3. An asterisk means a new designation. The designation of Bay (1992) has been followed for *Minuartia biflora* and *Phippsia algida* ssp. *algidiformis*.

Table 3. Percentage composition of the geographical distribution types according to Hultén (1958, 1964, 1971) in the West Greenland distribution types (WGDT). Circ: circumpolar; West: western distribution, i.e. main occurrence in North-America; East: eastern distribution, i.e. main occurrence in Eurasia; Amph: amphi-Atlantic; End: endemic to Greenland. The percentages from North and South Greenland are according to Bay (1992) and Feilberg (1984).

WGDT	Circ	West	East	Amph	End	Taxa, n
1	74	13	1	12		85
2	48	43	5	5		21
3	67	25	4	4		24
4	43	43		14		14
5	33	33		33		3
6	46	11	18	21	4	28
7	32	27	14	18	9	22
8	39	23	15	17	6	66
9	27	35	22	14	3	37
10	29	26	24	18	3	38
11	26	47	5	8	13	38
n taxa	171	98	39	51	14	376
W.Grl. %	46	26	10	14	4	
N.Grl. %	56	22	6	15	1	
S.Grl. %	46	23	18	12	2	

Table 4. List of taxa present in West Greenland. WGDT: West Greenland distribution type; BDT: biological distribution type according to Böcher (1975); GDT: geographical distribution type according to Hultén (1958, 1964, 1971).

Map no.	Taxon	WGDT	BDT	GDT
267.	*Agrostis hyperborea* Læst.	9a	B	East
177.	*Agrostis mertensii* Trin.	6d	LO	Circ
273.	*Agrostis stolonifera* L.	9c	B	Circ
287.	*Alchemilla alpina* L.	9d	LO	East
288.	*Alchemilla filicaulis* Bus.	9d	BO	East
232.	*Alchemilla glomerulans* Bus.	8d	LO	East
304.	*Alchemilla vestita* (Bus.) Raunk.	10a	BO	East
324.	*Alnus crispa* (Ait.) Pursh	10b	BC	West
201.	*Alopecurus aequalis* Sobol.	8a	B	Circ
109.	*Alopecurus alpinus* Sm.	3a	HA	Circ
274.	*Amerorchis rotundifolia* (Banks ex Pursh) Hult.	9c	BC	West
289.	*Andromeda polifolia* L.	9d	B	West
376.	*Anemone richardsonii* Hook.	11f	MA	West
202.	*Angelica archangelica* L. ssp. *norvegica* (Rupr.) Nordh.	8a	LO	East
196.	*Antennaria affinis* Fern.	7c	LC	End
359.	*Antennaria angustata* Greene	11c	MA	West
	Antennaria boecheriana A. E. Pors. (incl. in *A. canescens*)			

Table 4 – Continued

Map no.	Taxon	WGDT	BDT	GDT
74.	*Antennaria canescens* (Lge.) Malte	1e	LO	Amph
	Antennaria compacta Malte (incl. in *A. ekmaniana*)			
55.	*Antennaria ekmaniana* A. E. Pors.	1b	AC	West
342.	*Antennaria glabrata* (J. Vahl) Greene	11a	MA	West
219.	*Antennaria hansii* Kern.	8c	LO	End
220.	*Antennaria intermedia* (Rosenv.) M. P. Pors.	8c	LO*	End
363.	*Antennaria porsildii* Ekm.	11d	MA	Amph
	Antennaria sornborgeri Fern. (incl. in *A. canescens*)			
	Antennaria subcanescens Ostf. (incl. in *A. canescens*)			
168.	*Arabis alpina* L.	6c	LO	East
165.	*Arabis arenicola* (Richards.) Gel.	6b	MA	West
179.	*Arabis holboellii* Horn.	7a	LC	West
110.	*Arctagrostis latifolia* (R. Br.) Griseb.	3a	HA	Circ
377.	*Arctophila fulva* (Trin.) Anders.	11f	L	Circ
360.	*Arctostaphylos alpina* (L.) Spreng.	11c	L	Circ
368.	*Arctostaphylos uva-ursi* (L.) Spreng. ssp. *coactilis* (Fern. et Macbr.) Löve, Löve et Kapoor	11e	BC	West
56.	*Arenaria humifusa* Wbg.	1b	MA	West
57.	*Armeria scabra* Pall. ssp. *sibirica* (Turcz.) Hyl.	1b	AC	Circ
137.	*Arnica angustifolia* M. Vahl in Hornem.	4b	AC	Amph
151.	*Artemisia borealis* Pall.	6a	LC	Circ
325.	*Asplenium viride* Huds.	10b	BO	Circ
336.	*Athyrium distentifolium* Tausch ex Opiz	10c	LO	Circ
369.	*Atriplex longipes* Drej. ssp. *praecox* (Hülph.) Tures.	11e	B	East
152.	*Bartsia alpina* L.	6a	LO	East
305.	*Betula glandulosa* Michx.	10a	L	West
35.	*Betula nana* L.	1a	LC	East
326.	*Betula pubescens* Ehrh.	10b	BO	East
290.	*Botrychium boreale* Milde	9d	LO	Circ
233.	*Botrychium lanceolatum* (Gmel.) Ångstr.	8d	LO	Circ
192.	*Botrychium lunaria* (L.) Sw.	7b	B	Circ
347.	*Braya linearis* Rouy	11b	LC	Amph
348.	*Braya novae-angliae* (Rydb.) Th. Sør.	11b	LC	West
88.	*Braya purpurascens* (R. Br.) Bge.	2a	HA	Circ
89.	*Braya thorild-wulffii* Ostf.	2a	HA	West
203.	*Calamagrostis langsdorffii* (Link) Trin.	8a	BS	Circ
370.	*Calamagrostis lapponica* (Wbg.) Hartm. var. *groenlandica* Lge.	11e	MA*	End
180.	*Calamagrostis neglecta* (Ehrh.) Gaertn., Mey. et Scherb.	7a	B	Circ
253.	*Calamagrostis poluninii* Th. Sør.	8e	LC	End
138.	*Calamagrostis purpurascens* R. Br.	4b	AC	West
221.	*Callitriche anceps* Fern.	8c	B	West
268.	*Callitriche hamulata* Kütz.	9a	B	East
343.	*Callitriche hermaphroditica* L.	11a	B	Circ
204.	*Callitriche palustris* L.	8a	B	Circ
36.	*Campanula gieseckiana* Vest. in R. et S.	1a	L	Circ
21.	*Campanula uniflora* L.	1a	MA	Amph
1.	*Cardamine bellidifolia L.*	1a	A	Circ
68.	*Cardamine pratensis* L.	1c	L	Circ
337.	*Carex atrata* L.	10c	LO	Amph
100.	*Carex atrofusca* Schkuhr	2b	MA	Circ
181.	*Carex bicolor* All.	7a	L	Circ
37.	*Carex bigelowii* Torr.	1a	L	Circ
	Carex boecheriana Löve, Löve et Raymond (see *C. capillaris)*			
222.	*Carex brunnescens* (Pers.) Poir.	8c	L	Circ
223.	*Carex canescens* L.	8c	B	Circ
22.	*Carex capillaris* L. ssp. *fuscidula* (Krecz.) Löve et Löve	1a	L	Circ
349.	*Carex capillaris* L. ssp. *robustior* (Drej. ex Lge) Böch.	11b	LC	West
350.	*Carex capitata* L. ssp. *capitata*	11b	BC	Circ
148.	*Carex capitata* L. ssp. *arctogena* (H. Smith) Hiit.	5	L	West
234.	*Carex deflexa* Horn.	8d	BS	West
58.	*Carex glacialis* Mack.	1b	AC	Circ
75.	*Carex glareosa* Wbg.	1e	L	Circ
182.	*Carex gynocrates* Wormsk.	7a	LC	West
344.	*Carex holostoma* Drej.	11a	MA*	West
38.	*Carex lachenalii* Schkuhr	1a	L	Circ
169.	*Carex macloviana* D'Urv.	6c	L	Amph
327.	*Carex magellanica* Lam. ssp. *irrigua* (Wbg.) Hult.	10b	B	Circ
125.	*Carex marina* Dew.	3c	LC	Circ
18.	*Carex maritima* Gunn.	1a	A	Circ
197.	*Carex microglochin* Wbg.	7c	LC	Amph
120.	*Carex misandra* R. Br.	3b	AC	Circ
2.	*Carex nardina* Fr.	1a	AC	Amph
149.	*Carex norvegica* Retz.	5	LO	Amph
275.	*Carex praticola* Rydb.	9c	BC	West
153.	*Carex rariflora* (Wbg.) Sm.	6a	L	Circ
224.	*Carex rufina* Drej.	8c	LO	Amph*
139.	*Carex rupestris* All.	4b	AC	Circ
328.	*Carex salina* Wbg.	10b	B	Amph
69.	*Carex saxatilis* L.	1c	A	Circ
39.	*Carex scirpoidea* Michx.	1a	L	West
111.	*Carex stans* Drej.	3a	HA	Circ
338.	*Carex stylosa* C. A. Mey. var. *nigritella* (Drej.) Fern.	10c	BO	West
170.	*Carex subspathacea* Wormsk.	6c	L	Circ
59.	*Carex supina* Wbg. ssp. *spaniocarpa* (Steud.) Hult.	1b	LC	West
306.	*Carex trisperma* Dew.	10a	B	West
121.	*Carex ursina* Dew.	3b	MA	Circ
140.	*Cassiope tetragona* (L.) D. Don	4b	AC	Circ
307.	*Catabrosa aquatica* (L.) Beauv.	10a	B	East
	Cerastium alpinum L. (see *C. arcticum)*			
3.	*Cerastium arcticum* Lge. (incl. *C. alpinum)*	1a	A	Amph
378.	*Cerastium arvense* L.	11f	B	Circ
171.	*Cerastium cerastoides* (L.) Britton	6c	LO	East
308.	*Cerastium fontanum* Baumg. ssp. *scandicum* Gart.	10a	BO	East
225.	*Chamaenerion angustifolium* (L.) Scop.	8c	BS	Circ
23.	*Chamaenerion latifolium* (L.) Sweet	1a	A	Circ
76.	*Cochlearia groenlandica* L.	1e	A	Circ
101.	*Colpodium vahlianum* (Liebm.) Nevski	2b	HA	Circ
269.	*Comarum palustre* L.	9a	B	Circ
235.	*Coptis trifolia* (L.) Salisb.	8d	BS	West
254.	*Corallorrhiza trifida* Chât.	8e	BC	Circ
339.	*Cornus canadensis* L.	10c	BS	West
291.	*Cornus suecica* L.	9d	BO	Circ
24.	*Cystopteris fragilis* (L.) Bernh.	1a	( )	Circ

Table 4 – Continued

Map no.	Taxon	WGDT	BDT	GDT
309.	*Cystopteris montana* (Lam.) Desv.	10a	BC	Circ
292.	*Deschampsia alpina* (L.) R. et S.	9d	LO	East
310.	*Deschampsia flexuosa* (L.) Trin.	10a	BS	Amph
133.	*Deschampsia pumila* Ostf.	4a	HA	Circ
86.	*Diapensia lapponica* L. ssp. *lapponica*	1f	L	Amph
172.	*Diphasiastrum alpinum* (L.) Holub	6c	LO	Circ
236.	*Diphasiastrum complanatum* (L.) Holub	8d	BC	Circ
90.	*Draba adamsii* Led.	2a	HA	Circ
126.	*Draba alpina* L.	3c	MA	Circ
122.	*Draba arctica* J. Vahl	3b	AC	(Circ)
	Draba arctogena Ekm. (incl. in *D. norvegica*)			
183.	*Draba aurea* M.Vahl	7a	LC	West
91.	*Draba bellii* Holm	2a	HA	West
351.	*Draba cana* Rydb.	11b	LC	West
352.	*Draba cinerea* Adams	11b	LC	Circ
166.	*Draba crassifolia* Grah.	6b	MA	Amph
127.	*Draba fladnizensis* Wulf.	3c	A	East
141.	*Draba glabella* Pursh	4b	AC	Circ
	Draba groenlandica Ekm. (incl. in *D. arctica*)			
205.	*Draba incana* L.	8a	LO	Amph
60.	*Draba lactea* Adams	1b	A	Circ
40.	*Draba nivalis* Liljebl.	1a	A	Circ
193.	*Draba norvegica* Gunn.	7b	LO	Amph
92.	*Draba subcapitata* Simm.	2a	HA	Circ
329.	*Drosera rotundifolia* L.	10b	B	Circ
4.	*Dryas integrifolia* M. Vahl	1a	AC	West
293.	*Dryopteris assimilis* Walker	9d	BS	Amph
311.	*Dryopteris filix-mas* (L.) Schott.	10a	BS	Amph
130.	*Dryopteris fragrans* (L.) Schott.	3d	AC	Circ
364.	*Dupontia psilosantha* Rupr.	11d	MA	Circ
61.	*Eleocharis acicularis* (L.) R. et S.	1b	B	Circ
276.	*Eleocharis quinqueflora* (F. X. Hartm.) Schwarz	9c	B	Amph
365.	*Elymus hyperarcticus* (Polun.) Tzvel.	11d	AC	West
330.	*Elymus trachycaulus* (Link) Gould ssp. *virescens* (Lge.) Löve et Löve	10b	B	End
263.	*Elymus violaceus* (Horn.) Böch. et al. ex Feilberg	8f	LC	West
41.	*Empetrum nigrum* L. ssp. *hermaphroditum* (Hagerup) Böch.	1a	L	Circ
237.	*Epilobium anagallidifolium* Lam.	8d	LO	Circ
366.	*Epilobium arcticum* Sam.	11d	AC	Amph
238.	*Epilobium hornemannii* Rchb.	8d	LO	Circ
239.	*Epilobium lactiflorum* Hausskn.	8d	LO	Amph
206.	*Epilobium palustre* L.	8a	B	Circ
25.	*Equisetum arvense* L.	1a	B	Circ
207.	*Equisetum scirpoides* Michx.	8a	LC	Circ
240.	*Equisetum sylvaticum* L.	8d	BS	Circ
19.	*Equisetum variegatum* Schleich.	1a	L	Circ
	Erigeron borealis (Vierh.) Simm. (incl. in *E. uniflorus*)			
62.	*Erigeron compositus* Pursh	1b	AC	West
112.	*Erigeron eriocephalus* J. Vahl	3a	MA	Circ
142.	*Erigeron humilis* Grah.	4b	MA	West
226.	*Erigeron uniflorus* L.	8c	LO	East
42.	*Eriophorum angustifolium* Honck. ssp. *subarcticum* (V. Vassil.) Hult.	1a	B	Circ
26.	*Eriophorum scheuchzeri* Hoppe	1a	A	Circ
102.	*Eriophorum triste* (Th. Fr.) Hadac et Löve	2b	HA	Circ
154.	*Euphrasia frigida* Pugsl.	6a	L	Amph
103.	*Eutrema edwardsii* R. Br.	2b	HA	Circ
93.	*Festuca baffinensis* Polun.	2a	HA	West
5.	*Festuca brachyphylla* Schult. et Schult.	1a	A	Circ
184.	*Festuca groenlandica* (Schol.) Frederiksen	7a	L	(End)
104.	*Festuca hyperborea* Holmen ex Frederiksen	2b	HA	West
155.	*Festuca rubra* L. coll.	6a	L	Circ
208.	*Festuca saximontana* Rydb.	8a	B*	West
294.	*Festuca vivipara* (L.) Sm. var. *hirsuta* (Lge.) Schol.	9d	LO	Amph
277.	*Galium brandegei* A. Gray	9c	BC	West
340.	*Galium triflorum* Michx.	10c	BS	West
331.	*Gentiana amarella* L. ssp. *acuta* (Michx.) Hult.	10b	B	West
255.	*Gentiana aurea* L.	8e	B	East
278.	*Gentiana detonsa* Rottb.	9c	B	East
194.	*Gentiana nivalis* L.	7b	LO	East
353.	*Gentiana tenella* Rottb.	11b	L	East
241.	*Gnaphalium norvegicum* Gunn.	8d	LO	East
173.	*Gnaphalium supinum* L.	6c	LO	East
227.	*Gymnocarpium dryopteris* (L.) Newman	8c	BS	Circ
143.	*Halimolobus mollis* (Hook.) Rollins.	4b	LC	West
77.	*Harrimanella hypnoides* (L.) Coville	1e	LO	Amph
	Hieracium acranthophorum Om. (incl. in *H. rigorosum*)			
264.	*Hieracium alpinum* (L.) Backh.	8f	LO	East
242.	*Hieracium groenlandicum* (A.-T.) Almq.	8d	B	West
243.	*Hieracium hyparcticum* Almq.	8d	BS	End
	Hieracium ivigtutense (Almq.) Om. (incl. in *H. groenlandicum*)			
	Hieracium lividorubens Almq. (incl. in *H. hyparcticum*)			
295.	*Hieracium rigorosum* (Læst.) Almq.	9d	B	End
87.	*Hierochloë alpina* (Willd.) R. et S.	1f	AC	Circ
296.	*Hierochloë orthantha* Th. Sør.	9d	LO	West
27.	*Hippuris vulgaris* L.	1a	B	Circ
43.	*Honckenya peploides* (L.) Ehrh. var. *diffusa* (Horn.) Mattf.	1a	B	Circ
44.	*Huperzia selago* (L.) Bernh. ex Schrank et Mart.	1a	A	Circ
312.	*Isoëtes echinospora* Dur. ssp. *muricata* (Dur.) Löve et Löve	10a	B	West
256.	*Juncus alpinus* Vill. ssp. *nodulosus* (Wbg.) Lindm. (Wbg.) Lindm.	8e	B	Circ
156.	*Juncus arcticus* Willd.	6a	L	Circ
78.	*Juncus biglumis* L.	1e	A	Circ
134.	*Juncus castaneus* Sm.	4a	MA	Circ
313.	*Juncus filiformis* L.	10a	B	Circ
279.	*Juncus ranarius* Perr. et Song.	9c	B	Circ
209.	*Juncus subtilis* E. Mey.	8a	B	West
174.	*Juncus trifidus* L.	6c	LO	Amph
20.	*Juncus triglumis* L.	1a	A	Circ
210.	*Juniperus communis* L. ssp. *alpina* (Neilr.) Celak.	8a	B	Circ
28.	*Kobresia myosuroides* (Vill.) Fiori et Paol.	1a	LC	Circ
63.	*Kobresia simpliciuscula* (Wbg.) Mack.	1b	LC	Circ
29.	*Koenigia islandica* L.	1a	A	Circ

Table 4 – Continued

Map no.	Taxon	WGDT	BDT	GDT
371.	x *Ledodendron vanhoeffeni* (Abromeit) Dalgaard et Fredskild	11e	LC*	End
270.	*Ledum groenlandicum* Oed.	9a	B	West
135.	*Ledum palustre* L. ssp. *decumbens* (Ait.) Hult. (Ait.) Hult.	4a	LC	West
128.	*Lesquerella arctica* (Wormsk.) S. Wats.	3c	AC	West
244.	*Leucorchis albida* (L.) E. Mey. ssp. *straminea* (Fern.) Löve	8d	BO	West
257.	*Leymus arenarius* (L.) Hoechst.	8e	B	East
185.	*Leymus mollis* (Trin.) Pilger	7a	L	West
332.	*Ligusticum scoticum* L.	10b	BO	Amph
258.	*Limosella aquatica* L.	8e	B	Circ
245.	*Linnaea borealis* L. ssp. *americana* (Forb.) Hult.	8d	BS	West
246.	*Listera cordata* (L.) R. Br.	8d	BS	Amph
79.	*Loiseleuria procumbens* (L.) Desv.	1e	LO	Amph
280.	*Lomatogonium rotatum* (L.) Fr.	9c	L	West
123.	*Luzula arctica* Blytt	3b	AC	Circ
6.	*Luzula confusa* Lindeb.	1a	LO	Circ
354.	*Luzula groenlandica* Böch.	11b	LC	West
157.	*Luzula multiflora* (Retz.) Lej. ssp. *frigida* (Buch.) Krecz.	6a	L	Circ
333.	*Luzula multiflora* (Retz.) Lej. ssp. *multiflora*	10b	B	Amph*
228.	*Luzula parviflora* (Ehrh.) Desv.	8c	BS	Circ
150.	*Luzula spicata* (L.) DC.	5	LO	Circ
158.	*Lychnis alpina* L. ssp. *americana* (Fern.) Feilberg	6a	LO	West
159.	*Lycopodium annotinum* L. ssp. *alpestre* (Hartm.) Löve et Löve	6a	BS	Circ
314.	*Lycopodium clavatum* L. ssp. *monostachyon* (Grev. et Hook.) Sel.	10a	BS	Circ
229.	*Matricaria maritima* L. ssp. *borealis* (Hartm.) Löve et Löve	8c	B	Amph
144.	*Melandrium affine* J. Vahl coll.	4b	MA	West
105.	*Melandrium apetalum* (L.) Fenzl ssp. *arcticum* (Fr.) Hult.	2b	HA	Circ
129.	*Melandrium triflorum* (R. Br.) J. Vahl	3c	AC	West
281.	*Menyanthes trifoliata* L.	9c	B	Circ
72.	*Mertensia maritima* (L.) S. F. Gray	1d	L	West
80.	*Minuartia biflora* (L.) Sch. et Th.	1e	LO	Circ
297.	*Minartia groenlandica* (Retz.) Ostf.	9d	LO	West
94.	*Minuartia rossii* (R. Br.) Graebn.	2a	HA	West
7.	*Minuartia rubella* (Wbg.) Hiern	1a	A	Circ
113.	*Minuartia stricta* (Sw.) Hiern	3a	MA	Circ
211.	*Montia fontana* L. ssp. *fontana*	8a	B	Circ
271.	*Myriophyllum alterniflorum* DC.	9b	B	Amph
259.	*Myriophyllum spicatum* L. ssp. *exalbescens* (Fern.) Hult.	8e	B	West
315.	*Nardus stricta* L.	10a	BO	East
361.	*Orthilia secunda* (L.) House ssp. *obtusata* (Turcz.) Böch.	11c	BC	West
8.	*Oxyria digyna* (L.) Hill	1a	A	Circ
9.	*Papaver radicatum* Rotth. coll.	1a	A	Circ
265.	*Parnassia kotzebuei* Cham. et Schlecht.	8f	L	West
45.	*Pedicularis flammea* L.	1a	L	West
379.	*Pedicularis groenlandica* Retz.	11f	BO	West
145.	*Pedicularis hirsuta* L.	4b	AC	Amph
282.	*Pedicularis labradorica* Wirsing	9c	LC	West
345.	*Pedicularis lanata* Cham. et Schlecht.	11a	AC	West
160.	*Pedicularis lapponica* L.	6a	LC	Circ
316.	*Phegopteris connectilis* (Michx.) Watt	10a	BS	Circ
64.	*Phippsia algida* (Sol.) R. Br.	1b	A	Circ
106.	*Phippsia algida* (Sol.) R. Br. ssp. *algidiformis* (H. Sm.) L. & L.	2b	HA	East
247.	*Phleum commutatum* Gaud.	8d	LO	Circ
81.	*Phyllodoce coerulea* (L.) Bab.	1e	LO	Circ
186.	*Pinguicula vulgaris* L.	7a	B	Amph
213.	*Plantago maritima* L. ssp. *borealis* (Lge) Blytt & Dahl	8a	B	Circ
248.	*Platanthera hyperborea* (L.) Lindl.	8d	BS	West
95.	*Poa abbreviata* R. Br.	2a	HA	Amph
161.	*Poa alpina* L.	6a	LO	Amph
	Poa arctica R. Br. (incl. in *P. pratensis*)			
212.	*Poa flexuosa* Sm.	8a	L*	Amph
30.	*Poa glauca* M. Vahl	1a	A	Circ
96.	*Poa hartzii* Gandoger	2a	AC	West
317.	*Poa nemoralis* L.	10a	BS	Circ
10.	*Poa pratensis* L. coll.	1a	B	Circ
97.	*Poa pratensis* L. var. *colpodea* (Th. Fr.) Schol.	2a	HA	Circ
11.	*Polygonum viviparum* L.	1a	A	Circ
249.	*Polystichum lonchitis* (L.) Roth	8d	LO	Amph
283.	*Potamogeton alpinus* Balb. ssp. *tenuifolius* (Raf.) Hult.	9c	B	West
198.	*Potamogeton filiformis* Pers.	7c	B	Amph
284.	*Potamogeton gramineus* L.	9c	B	Circ
162.	*Potamogeton pusillus* L. ssp. *groenlandicus* (Hagstr.) Böch.	6a	B	End
163.	*Potentilla crantzii* (Cr.) G. Beck	6a	LO	Amph
187.	*Potentilla egedii* Wormsk.	7a	L	West
146.	*Potentilla hookeriana* Lehm.	4b	AC	Circ
114.	*Potentilla hyparctica* Malte	3a	HA	Circ
46.	*Potentilla nivea* L. emend. Hult.	1a	AC	Circ
124.	*Potentilla pulchella* R. Br.	3b	HA	Amph
266.	*Potentilla ranunculus* Lge.	8f	BO	West
98.	*Potentilla rubricaulis* Lehm.	2a	HA	West
218.	*Potentilla tridentata* Sol. in Ait.	8b	B	West
115.	*Potentilla vahliana* Lehm.	3a	AC	West
285.	*Primula egaliksensis* Wormsk.	9c	LC	West
355.	*Primula stricta* Horn.	11c	LC	West
99.	*Puccinellia andersonii* Swall.	2a	HA	West
116.	*Puccinellia angustata* (R.Br.) Rand. et Redf.	3a	HA	Circ
214.	*Puccinellia coarctata* Fern. et Weath.	8a	L	Amph
356.	*Puccinellia deschampsioides* Th. Sør.	11b	LC	West
357.	*Puccinellia groenlandica* Th. Sør.	11b	LC*	End
362.	*Puccinellia langeana* (Berlin) Th. Sør.	11c	MA	West
318.	*Puccinellia maritima* (Huds.) Parl.	10a	B	East
82.	*Puccinellia phryganodes* (Trin.) Schribn. et Merr.	1e	A	Circ
367.	*Puccinellia rosenkrantzii* Th. Sør.	11d	MA	End
117.	*Puccinellia vaginata* (Lge.) Fern. et Weath.	3a	MA	West
47.	*Pyrola grandiflora* Rad.	1a	AC	Circ
230.	*Pyrola minor* L.	8c	BS	Circ
298.	*Ranunculus acris* L.	9d	B	Circ
131.	*Ranunculus affinis* R. Br.	3d	LC	West
70.	*Ranunculus confervoides* Fr.	1c	L	Circ
372.	*Ranunculus cymbalaria* Pursh	11e	B	West
31.	*Ranunculus hyperboreus* Rottb.	1a	A	Circ
346.	*Ranunculus lapponicus* L.	11a	LC	Circ
118.	*Ranunculus nivalis* L.	3a	MA	Circ
48.	*Ranunculus pygmaeus* Wbg.	1a	MA	Circ

Table 4 – Continued

Map no.	Taxon	WGDT	BDT	GDT
260.	*Ranunculus reptans* L.	8e	B	Circ
107.	*Ranunculus sulphureus* Sol. in Phipps	2b	HA	Circ
319.	*Rhinanthus minor* L.	10a	B	Amph
231.	*Rhodiola rosea* L.	8c	LO	East
49.	*Rhododendron lapponicum* (L.) Wbg.	1a	AC	West
286.	*Rorippa islandica* (Oed.) Borb.	9c	B	Circ
341.	*Rubus chamaemorus* L.	10c	B	Circ
199.	*Rumex acetosella* L. coll.	7c	B	Circ
65.	*Sagina caespitosa* (J. Vahl) Lge	1b	MA	West
83.	*Sagina intermedia* Fenzl	1e	A	Circ
334.	*Sagina procumbens* L.	10b	B	East
250.	*Sagina saginoides* (L.) Karst.	8d	B	East
119.	*Salix arctica* Pall.	3a	HA	Circ
188.	*Salix arctophila* Cockerell	7a	L	West
50.	*Salix glauca* L. coll.	1a	L	Circ
119.	*Salix herbacea* L.	1a	LO	Amph
335.	*Salix uva-ursi* Pursh	10b	LO	West
71.	*Saxifraga aizoides* L.	1c	L	Amph.
12.	*Saxifraga caespitosa* L.	1a	A	Circ
13.	*Saxifraga cernua* L.	1a	A	Circ
32.	*Saxifraga foliolosa* R. Br.	1a	HA	Circ
	Saxifraga hyperborea R. Br. (incl. in *S. rivularis*)			
14.	*Saxifraga nivalis* L.	1a	A	Circ
15.	*Saxifraga oppositifolia* L.	1a	A	Circ
52.	*Saxifraga paniculata* Mill.	1a	L	West
16.	*Saxifraga rivularis* L.	1a	L	Circ
299.	*Saxifraga stellaris* L.	9d	LO	East
84.	*Saxifraga tenuis* (Wbg.) H. Smith	1e	A	Circ
147.	*Saxifraga tricuspidata* Rottb.	4b	AC	West
178.	*Scirpus caespitosus* L.	6d	BO	Circ
320.	*Sedum annuum* L.	10a	LO	East
189.	*Sedum villosum* L.	7a	B	East
300.	*Selaginella selaginoides* (L.) Link	9d	LO	Circ
175.	*Sibbaldia procumbens* L.	6c	LO	Circ
33.	*Silene acaulis* (L.) Jacq.	1a	A	Circ
373.	*Sisyrinchium groenlandicum* Böch.	11e	BC	End
215.	*Sparganium angustifolium* Michx.	8a	B	Amph
216.	*Sparganium hyperboreum* Læst. ex Beurl.	8a	L	Circ
374.	*Spergularia canadensis* (Pers.) G. Don	11e	B*	West
251.	*Stellaria calycantha* (Led.) Bong.	8d	BS	West
17.	*Stellaria longipes* Goldie coll. (sepals glabrous: »A«)	1a	( )	( )
136.	*Stellaria longipes* Goldie coll. (sepals ciliate: »B«)	4a	( )	( )
	Stellaria spp.: see Taxonomical remarks			
85.	*Stellaria humifusa* Rottb.	1e	A	Circ
261.	*Subularia aquatica* L.	8e	B	Amph
	Taraxacum amphiphron Böch. (incl. in *T. croceum*)			
176.	*Taraxacum croceum* Dahlst.	6c	LO	East
167.	*Taraxacum lacerum* Greene	6b	LC	West
108.	*Taraxacum phymatocarpum* J. Vahl	2b	HA	West
	Taraxacum umbrinum Dahlst. (incl. in *T. lacerum*)			
190.	*Thalictrum alpinum* L.	7a	LO	Circ
301.	*Thymus praecox* Opiz ssp. *arcticus* (E. Durand) Jalas	9d	BO	East
132.	*Tofieldia coccinea* Richards.	3d	AC	West
53.	*Tofieldia pusilla* (Michx.) Pers.	1a	L	Circ
191.	*Triglochin palustre* L.	7a	B	Circ
34.	*Trisetum spicatum* (L.) Richt.	1a	A	Circ
217.	*Trisetum triflorum* (Bigel.) Löve et Löve	8a	L	Amph
262.	*Utricularia intermedia* Hayne	8e	B	Circ
200.	*Utricularia minor* L.	7c	B	Circ
358.	*Utricularia ochroleuca* R. Hartm.	11b	B*	Circ
272.	*Vaccinium oxycoccus* L. ssp. *microphyllum* (Lange) Feilberg	9b	B	West
54.	*Vaccinium uliginosum* L.	1a	L	Circ
73.	*Vaccinium vitis-idaea* L. ssp. *minus* (Lodd.) Hult.	1d	B	Circ
302.	*Vahlodea atropurpurea* (Wbg.) Fr.	9d	LO	Amph
195.	*Veronica alpina* L.	7b	B	East
252.	*Veronica fruticans* Jacq.	8d	LO	East
303.	*Veronica wormskjoldii* R. et S.	9d	LO	West
321.	*Viola labradorica* Schrank	10a	B	West
322.	*Viola palustris* L.	10a	BO	East
323.	*Viola selkirkii* Pursh	10a	BC	Circ
66.	*Woodsia alpina* (Bolt.) S. F. Gray	1b	L	Amph
67.	*Woodsia glabella* R. Br.	1b	A	Circ
164.	*Woodsia ilvensis* (L.) R. Br.	6a	L	Circ
375.	*Zostera marina* L.	11e	B	Circ

6.2. Species diversity in Greenland

To illustrate the species diversity Fig. 5 gives the number of taxa collected at 15 localities (three of which are G.B.S. loc.) with more than 300 collections, and at 14 well-investigated G.B.S. localities. With two exceptions the lowest numbers are seen along the outer coast, along the southern coast of Disko Bugt, and in the northernmost part of West Greenland. Two exceptions are Sisimiut, partly situated in a protected valley behind high coastal mountains, and Qeqertarsuaq with its extreme richness in habitats, including the homothermic springs. When passing the boundary between the low and high arctic zones at c. 70°N the number decreases below 150. By comparison, in Northwest Greenland the richest locality in Melville Bugt (Tugtuligssuaq at 75°N) has 83 taxa, in the Thule district (Thule, c. 76°30'N) 113, in the interior North Greenland (Brainard Sund, 83°N) 94, and in coastal North Greenland (Station Nord, c. 81°30'N) only 39 taxa (Bay 1992). In the southern part of the high arctic zone in East Greenland, Zackenberg (74°28'N), so far the most diverse Greenland locality north of 70°N, has 149 taxa, about the same number as a rich inland locality in low arctic Southeast Greenland, viz. Dronning Maries Dal (c. 63°30'N, Fredskild & Bay 1993).

In South Greenland the number of taxa in the subarctic inland (in areas of birch forest in the lowland between high alpine mountains) is 267 at the head of the Tunugdliarfik fiord (61°N). At Kangerluarsuk half way out in the fiord at the transition between the oceanic and the sub oceanic, low arctic zone 185 taxa are found, and even further out, on the north end of the outer coast isle Sermersooq (c. 60°30'N) 120 (Feilberg 1984). As mentioned above the species delimitation differs slightly between the present paper and those quoted.

6.3. Delimitation of floristic provinces and districts

Study of the 379 maps reveals that there are coinciding distribution limits for different species in West Greenland (Fig. 6). These are either northern or southern limits. The number to which a line marks the absolute limit is given, which, of course, means that most but not all taxa just reach the limit. Besides, the number of taxa with a disjunct distribution, to which the line in question marks a limit, is given in parenthesis. No less than 40 taxa have their West Greenland northern limits at the lines at 71°-72°N and at 73°N, to reappear in Northwest Greenland.

Of all, the line cutting Disko-Nuussuaq is the most marked, showing the absolute northern limit of 62 and southern limit of 18 taxa. Thus, this line is considered the floristic border between low and high arctic West Greenland. The second most important is the line through Maniitsoq ice cap at 66°N, which marks the absolute limit of 16 and 34 taxa, resp.

However, these east-west running lines do not inform on the distribution in an east-west direction, mainly reflecting the degree of continentality. The map, Fig. 7, combines the lines of northern and southern limits with the most pronounced lines when looking at maps of plants with a clear oceanic or continental distribution. The most pronounced of the latter lines in the low arctic runs almost parallel with the coast line until the south boundary of Disko Bugt. From here, it differs somewhat from the first map of the Greenland floristic provinces and districts, published in Böcher & al. (1957) and further discussed in Böcher & al. (1959). The relevant part of this map, which was brought unchanged in all editions of the Flora, is shown here in Fig. 8. The new course of the line is mainly a result of many years G.B.S. investigations which may also account for the displacement of the boundary between low and high arctic northwards on Disko. Böcher & al. (1957) includes the eastern half of the area Nuussuaq-Svartenhuk in the district CWn, thus including it in the low arctic zone. From a climatic point of view this seems corroborated by the northernmost occurrences of willow scrubs on isolated, favoured places at the very head of the fiords on both sides of Svartenhuk. However, from a phytogeographical point of view the Disko-Nuussuaq line seems more obvious, as it marks the southern limit of 18 high arctic taxa whereas neither of the three lines north of it (Fig. 6) is southern limit to any taxon. Since many of the high arctic taxa, e.g. of Type 2, only occur on north Disko-west Nuussuaq, and, on the contrary, many low arctic taxa, e.g. of Type 7 and 8, towards their northern limit only occur in the most eastern part of the region 70°-73°N, it is suggested to divide the district NWs into an outer part: NWso, and an inner part: NWsi.

Fig. 6. The most important north-south distributional limits in West Greenland. Numbers below a line indicate northern limits, those above a line are southern limits. Numbers in parenthesis indicate that the line is a marked West Greenland limit to a taxon with a disjunct distribution in Greenland.

Fig. 7. Map of West Greenland with the new boundaries between floristic provinces (full lines) and districts (dashed lines). The line between NWso/NWsi, and SWn/CWn is considered the phytogeographical boundary between high and low arctic Greenland.

Fig. 8. The floristic provinces and districts according to Böcher & al. 1959 and all editions of Grønlands Flora/Flora of Greenland.

Only slight alterations have been made in the boundaries between the provinces and districts in the southern part of West Greenland. The boundary between CWm and CWn has been moved to the south, and the southern part of the province border SW-CW is moved closer to the outer coast.

According to Feilberg (1984, Fig. 11) the most distinctive floristic boundary in the southwestern part of Greenland lies at 62°20'N, which he considers the northern border of the floristic province South Greenland. This I accept here, and consequently the coastal, floristic province Southwest Greenland is only subdivided into SWn and SWs. Judging from the numbers of taxa with northern or southern limits given in Feilberg (1985, Fig. 12), the boundary slightly more to the south between Kvanefjord and Neria districts at c. 61°30'-62°N is equally important. However, among the 15 taxa to which this is the northern limit, four are introduced, and two are found north of the border, viz. *Gentiana amarella* at the head of Godthåbsfjord (map 331), and *Carex trisperma* at 62°21'N (map 306).

7. Acknowledgements

Many colleagues and students have contributed many collections to the Greenland herbarium during the 22 summers Greenland Botanical Survey has worked in West Greenland. In the 1960s Greenland Geological Survey supported the field work logistically, not in the least by helicopter transport to remote places. In the 1970s and 1980s the Arctic Station on Disko has been especially important. Its cutter "Porsild" has been invaluable during the many excursions in the Disko Bugt area and in the long fiords to the south of it. In this connection I would like especially to thank the former leaders

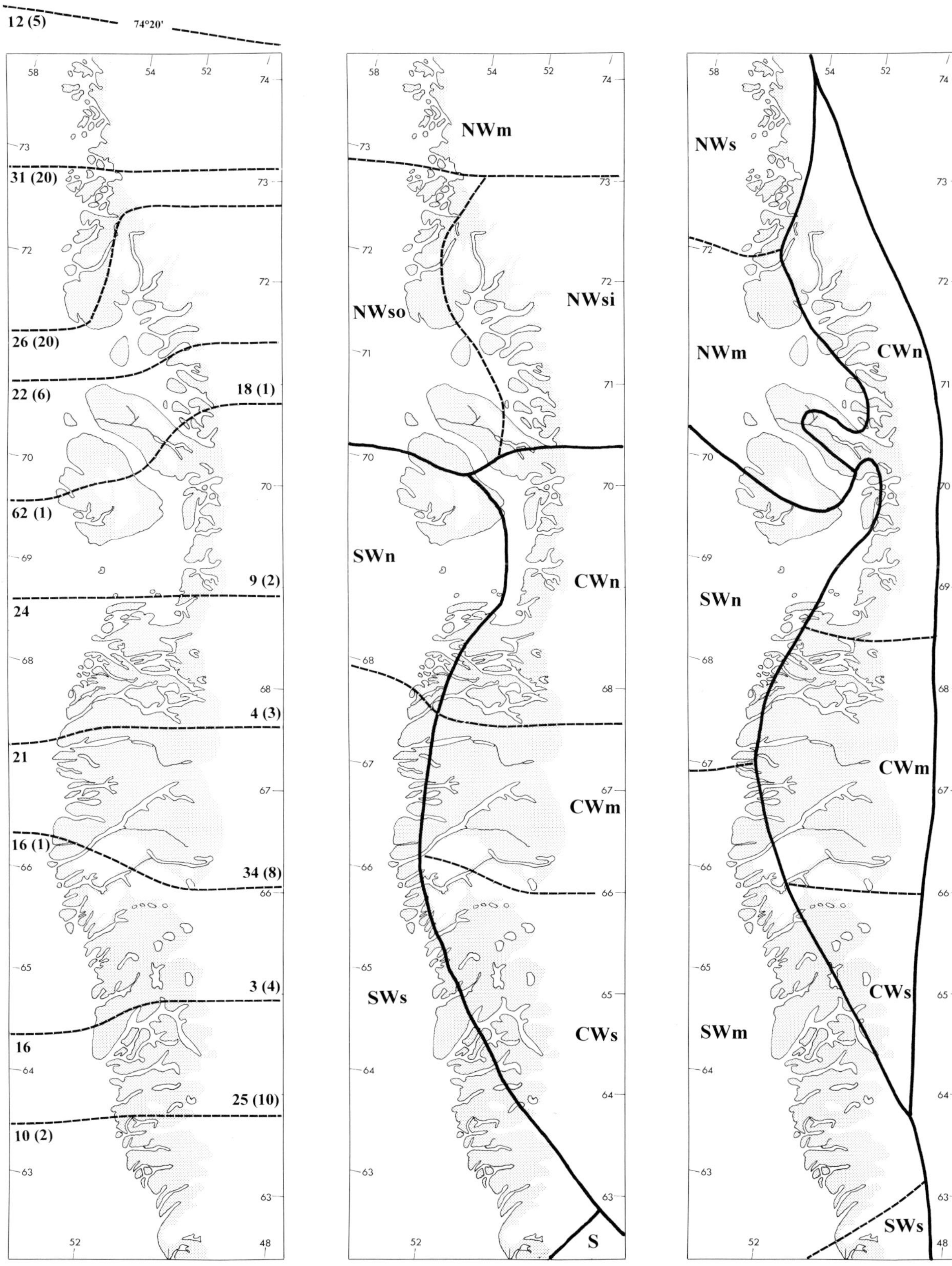

Fig. 6.

Fig. 7.

Fig. 8.

of the station Jon Feilberg and Vilhelm Dalgaard with whom I spent several summers.

Other large and important collections have been given to the Botanical Museum by many lovers of Greenland natural history, amateurs as well as professionals. In the recent decades the largest collections from W.Greenland were made by Villy Blom, David Boertmann, Eric Steen Hansen, Inger Hauge, Sune Holt, and Beate and Morten Strandberg. Birthe Hammer has carefully undertaken the major task of labelling all the material brought home. My best thanks are due both to the persons mentioned and to the other contributors. Especially, I would like to thank Christian Bay for the many summers we have worked together in Greenland, and for his comments on the manuscript.

Postscript

Aiken & al. (1995) describe a new species from the High Arctic: *Festuca edlundiae,* associated with *F. hyperborea* Holmen ex Frederiksen. A specimen from Umiviup kangerdlua, Svartenhuk (800 m a.s.l., 71°36 – 38'N, 54°10'W, leg K. Jakobsen Aug. 21, 1950), plotted on map 104 as *F. hyperborea,* has been redetermined by S. Aiken in 1995 as *F. edlundiae.* The locality is the southernmost in Greenland.

8. References

Ahokas, H. & Fredskild, B. 1991. Coexistence and hybridization of Leymus mollis and L. arenarius in Greenland, and demarcation of the species by endospermal prolamins, leymins. – Nord. J. Bot. 11: 385-392.

Aiken, S.G., Consaul, L.L. and Lefkovitch, L.P. 1995. Festuca edlundiae (Poaceae), a High Arctic, New Species Compared Enzymatically and Morphologically with Similar Festuca Species. – Systematic Botany 20.3: 374-392.

Anonymous 1979-1991. Grønland. Kalaallit Nunaat. – Årbog. Statsministeriet. Grønlandsafdelingen. Copenhagen.

Bay, C. 1992. A phytogeographical study of the vascular plants of northern Greenland – north of 74° northern latitude. – Meddr Grønland, Biosci. 36: 102 pp.

Böcher, J. & Fredskild, B. 1993. Plant and arthropod remains from the palaeo-Eskimo site on Qeqertasussuk, West Greenland. – Meddr Grønland, Geosci. 30: 35 pp.

Böcher, T. W. 1949. Climate, soil, and lakes in continental West Greenland in relation to plant life. – Meddr Grønland 147 (2): 63 pp.

– 1951. Distributions of plants in the circumpolar area in relation to ecological and historical factors. – J. Ecol. 39: 376-395.

– 1952. A study of the circumpolar Carex Heleonastes-amblyorhyncha complex. – Acta Arctica 5: 32 pp.

– 1954. Oceanic and continental vegetational complexes in Southwest Greenland. – Meddr Grønland 148 (1): 336 pp.

– 1959. Floristic and ecological studies in Middle West Greenland. – Meddr Grønland 156 (5): 68 pp.

– 1963. Phytogeography of Middle West Greenland. – Meddr Grønland 148 (3): 289 pp.

– 1966. Experimental and cytological studies on plant species. IX. Some arctic and montane Crucifers. – Biol. Skr. Dan. Vid. Selsk. 14 (7): 74 pp.

– 1975. Det grønne Grønland. – København, 256 pp.

– 1977. Cerastium alpinum and C. arcticum, a mature polyploid complex. – Bot. Notiser 130: 303-309.

– , Holmen, K. & Jakobsen, K. 1957. Grønlands Flora. – Copenhagen, 313 pp.

– , Holmen, K. & Jakobsen, K. 1959. A synoptical study of the Greenland flora. – Meddr Grønland 163(1): 32 pp.

– , Holmen, K. & Jakobsen, K. 1968. The Flora of Greenland (2. edition) – Copenhagen, 312 pp.

– , Fredskild, B., Holmen, K. & Jakobsen, K. 1978. Grønlands Flora (3. edition) – København, 327 pp.

Dalgaard, V. & Fredskild, B. 1993. x Ledodendron vanhoeffeni (syn.: Rhododendron vanhoeffeni) refound in Greenland. – Nord. J. Bot. 13: 253-255.

Dijkmans, J. W. A. & Törnqvist, T. E. 1991. Modern periglacial eolian deposits and landforms in the Søndre Strømfjord area, West Greenland and their palaeoenvironmental implications. – Meddr Grønland, Geosci. 25, 39 pp.

Eisner, W. R, Törnqvist, T. E., Koster, E. A., Bennike, O. & van Leeuwen, J. F. N. 1995. Paleoecological studies of a Holocene lacustrine record from the Kangerlussuaq (Søndre Strømfjord) region of West Greenland. – Quaternary Research 43,55-66.

Escher, A. & Watt, W. S. (ed.) 1976. Geology of Greenland. – Copenhagen, 603 pp.

Feilberg, J. 1984. A phytogeographical study of South Greenland. Vascular plants. – Meddr Grønland, Biosci. 15: 70 pp.

– 1985. Grønlands varme kilder – naturens egne mistbænke. – forskning i Grønland/tusaat 85,2: 10-22.

Frederiksen, S. 1981. Festuca vivipara (Poaceae) in the North Atlantic area. Nord. J. Bot. 1: 277-292.

– 1982. Festuca brachyphylla, F. saximontana and related species in North America. – Nord. J. Bot. 2: 525-536.

Fredskild, B. 1961. Floristic and ecological studies near Jakobshavn, West Greenland. Meddr Grønland 163 (4): 82 pp.

– 1973. Studies in the vegetational history of Greenland. Palaeobotanical investigations of some Holocene lake and bog deposits. – Meddr Grønland 198 (4): 247 pp.

– 1978. Palaeobotanical investigations of some peat deposits of Norse age at Qagssiarssuk, South Greenland. – Meddr Grønland 204 (5): 41 pp.

– 1984a. The Holocene vegetational development of the Godthåbsfjord area, West Greenland. – Meddr Grønland, Geosci. 10: 28 pp.

– 1984b. Distribution and occurrence of Onagraceae in Greenland. – Nord. J. Bot. 4: 475-480.

– 1985. The Holocene vegetational development of Tugtuligssuaq and Qeqertat, Northwest Greenland. – Meddr Grønland, Geosci. 14: 20 pp.

– 1991. The genus Betula in Greenland – Holocene history, present distribution and synecology. – Nord. J. Bot. 11: 393-412.

– 1992. The Greenland limnophytes – their present distribution and Holocene history. – Acta Bot. Fennica 144: 93-113.

– & Bay, C. (ed.) 1993. Grønlands Botaniske Undersøgelse. Greenland Botanical Survey. 1992: 51 pp. – Botanisk Museum, København.

– & Holt, S. 1993. The West Greenland "Greens" – favourite caribou summer grazing areas and Late Holocene climatic changes. – Geogr. Tidsskr. 93:30-38.

– & Ødum, S. 1990. The Greenland Mountain birch zone, an introduction. – Meddr Grønland, Biosci. 33: 3-7.

Funder, S. 1989. Quaternary geology of West Greenland. In: Fulton, R.J. (ed.) Quaternary Geology of Canada and Greenland. – Geological Survey of Canada: 749-756.

Gjærevoll, O. & Ryvarden, L. 1977. Botanical Investigations on J.A.D. Jensens Nunatakker in Greenland. – K. norske Vidensk. Selsk. Skr. 4: 1-40.

Halliday, G., Kliim-Nielsen, L. & Smart, I. H. M. 1974. Studies on the flora of the north Blosseville Kyst and on the hot springs of Greenland. – Meddr Grønland 199.2: 49 pp.

Hansen, K. 1967. The general limnology of arctic lakes as illustrated by examples from Greenland. – Meddr Grønland 178 (3): 77 pp.

– 1970. Geological and geographical investigations in Kong Frederik IX's Land, morphology, sediments, periglacial processes and salt lakes. – Meddr Grønland 188 (4): 77 pp.

Hansen, Kj. 1969. Analyses of soil profiles in dwarf-shrub vegetation in South Greenland. – Meddr Grønland 178 (5): 33 pp.

Holowaychuk, N. & Everett, K. R. 1972. Soils of the Tasersiaq area, Greenland. – Meddr Grønland 188 (6): 35 pp.

Hultén, E. 1937. Outline of the history of Arctic and Boreal Biota during the Quaternary Period. – Stockholm: 168 pp.

– 1958. The amphi-Atlantic plants and their phytogeographical connections. – K. Svenska Vetensk. Akad. Handl. 4,7,1: 340 pp.

– 1964. The circumpolar plants. I. – K. Svenska Vetensk. Akad. Handl. 4,8,5: 280 pp.

– 1971. The circumpolar plants. II. – K. Svenska Vetensk. Akad. Handl. 4,13,1: 463 pp.

Ingólfsson, Ó., Frich, P., Funder, S. & Humlum, O. 1990. Paleoclimatic implications of an early Holocene glacier advance on Disko Island, West Greenland. – Boreas 19: 297-311.

Iversen, J. 1940. Blütenbiologische Studien. I. Dimorphie und Monomorphie bei Armeria. – Biol. Skr. Vid. Selsk. 15 (8): 1-39.

– 1954. Origin of the flora of Western Greenland in the light of pollen analysis. – Oikos 4, 2: 85-103.

Jakobsen, B. H. 1988. Soil formation on the peninsula Tugtuligssuaq, Melville Bay, North West Greenland. – Geogr. Tidsskr. 88: 86-93.

Jørgensen, C. A., Sørensen, T. & Westergaard, M. 1958. The flowering plants of Greenland. A taxonomical and cytological survey. – Biol. Skr. Vid. Selsk. 9 (4): 172 pp.

Kelly, M. & Funder, S. 1974. The pollen stratigraphy of late Quaternary lake sediments of South-West Greenland. – Rapp. Grønlands geol. Unders. 64: 26 pp.

Lange, J. 1880. Conspectus Florae Groenlandicae. – Meddr Grønland 3 (1): 229 pp.

Mogensen, G. S. 1988. Taxonomy and distributrion of Greenland mosses. I. Encalypta brevipes, E. mutica, E. longicolla, and Amblyodon dealbatus. – Beiheft zur Nova Hedwigia 90: 387-397.

Pedersen, A. 1972. Adventitious plants and cultivated plants in Greenland. – Meddr Grønland 178 (7): 99 pp.

Petersen, P.M. 1981. Variation of the population structure of Polygonum viviparum L. in relation to certain environmental conditions. – Meddr Grønland, Biosci. 4:19 pp.

Philipp, M. 1972. The Stellaria longipes group in N.W.Greenland. Cytological and morphological investigations. – Bot. Tidsskr. 67: 64-75.

– 1978. Vegetation of a snow bed at Godhavn, West Greenland. – Holarct. Ecol. 1: 46-53.

Polunin, N. 1943. Contributions to the Flora and Phytogeography of south-western Greenland: an enumeration of the vascular plants, with critical notes. – Journ. Linn. Soc.-Botany. 52: 349-406.

Porsild, A. E. 1926. Contributions to the flora of West Greenland at 70°-71°45´N. Lat. Meddr Grønland 58 (2): 157-196.

– 1965. The genus Antennaria in eastern Arctic and Subarctic America. – Bot. Tidsskr. 61: 22-55.

– & Cody, W. J. 1979. Vascular plants of continental Northwest Territories, Canada. – Nat. Mus. of Canada, Ottawa: 667 pp.

Porsild, M. P. 1912. Vascular Plants of West Greenland between 71° and 73°N. Lat. Meddr Grønland 50 (7): 349-389.

– 1920. The flora of Disko island and the adjacent coast of West Greenland. – Meddr Grønland 58 (1): 156 pp.

– 1935. On some herbaria from Greenland and Labrador collected by the Moravian Brethren. – Meddr Grønland 93 (3): 84-94.

Putnins, P. 1971. The Climate of Greenland. – In: Orvig, S. (ed.): Climates of the Polar Regions. World Survey of Climatology 14: 1-128.

Raymond, M. 1949. Notes sur le genre Carex II. La valeur taxonomique de C. arctogena. – Contrib. Inst. bot. Univ. Montréal 64: 37-41.

Røen, U. I. 1962. Studies on freshwater Entomostraca in Greenland. II. – Meddr Grønland 170 (2): 249 pp.

Scoggan, H. J. 1978-79. The Flora of Canada 1-4. – Nat. Mus. of Canada, Ottawa: 1711 pp.

Stäblein, G. 1977. Arktische Böden West-Grönlands: Pedovarianz in Abhängigkeit vom geoökologischen Milieu. – Polarforschung 47: 11-25.

Sørensen, T. 1943. The Flora of Melville Bugt. – Meddr Grønland 124 (5): 70 pp.

– 1953. A revision of the Greenland species of Puccinellia Parl. – Meddr Grønland 136 (3): 179 pp.

– 1954. New species of Hierochloë, Calamagrostis, and Braya. – Meddr Grønland 136 (8): 24 pp.

Tuhkanen, S. 1980. Climatic parameters and indices in plant geography. – Acta Phytogeographica Suecica 67, 110. pp.

Tutin, T. G. (ed.) 1964. Flora Europaea. I. – Cambridge University Press: 464 pp.

Vestergaard, P. 1978. Studies in vegetation and soil of coastal salt marshes in the Disko area, West Greenland. – Meddr Grønland 204 (2): 51 pp.

Warming, E. 1888. Om Grønlands Vegetation. Meddr Grønland 12: 245 pp.

Weidick, A. 1968. Observarions on some Holocene glacier fluctuations in West Greenland. – Meddr Grønland 165 (6): 203 pp.

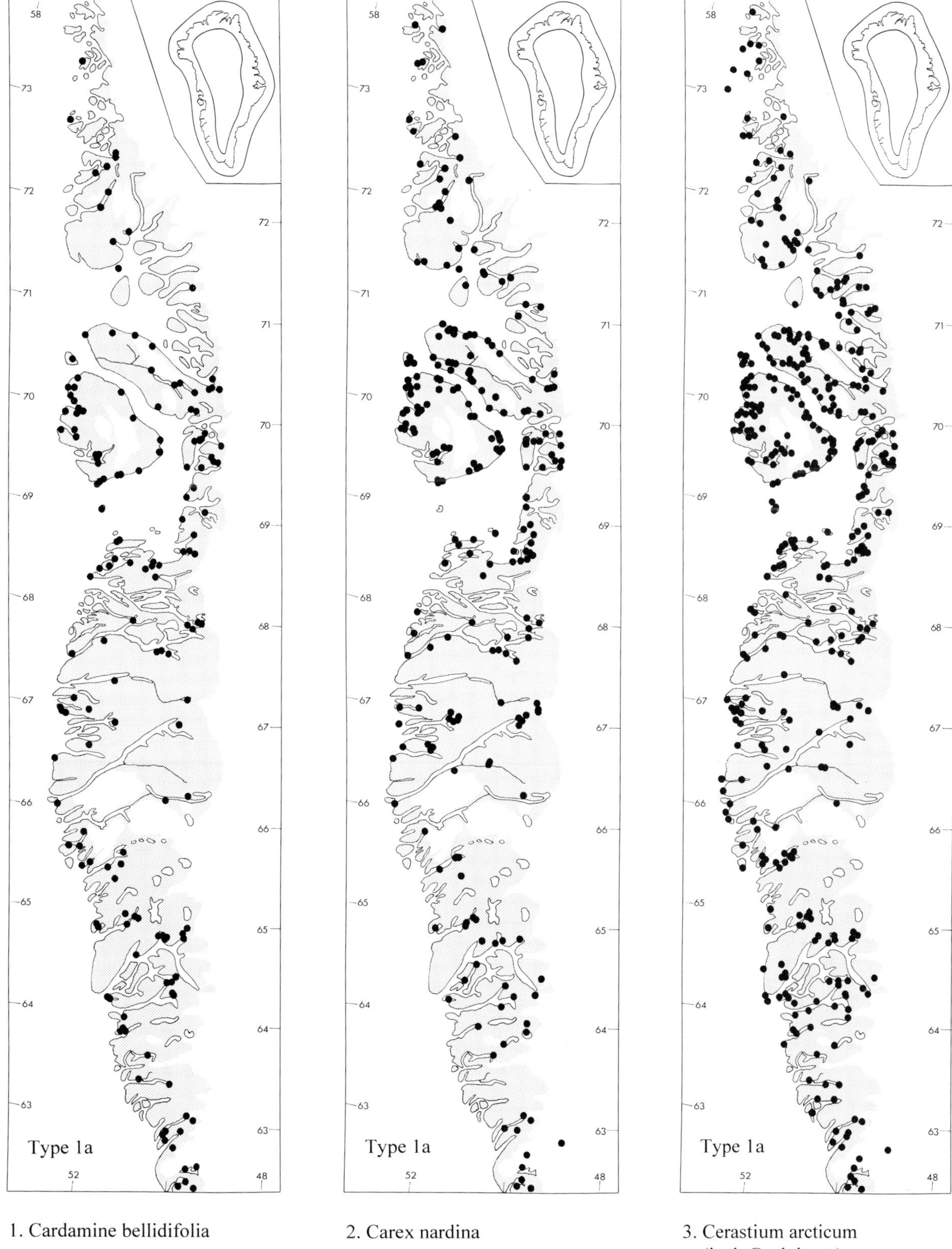

1. Cardamine bellidifolia

2. Carex nardina

3. Cerastium arcticum (incl. C. alpinum)

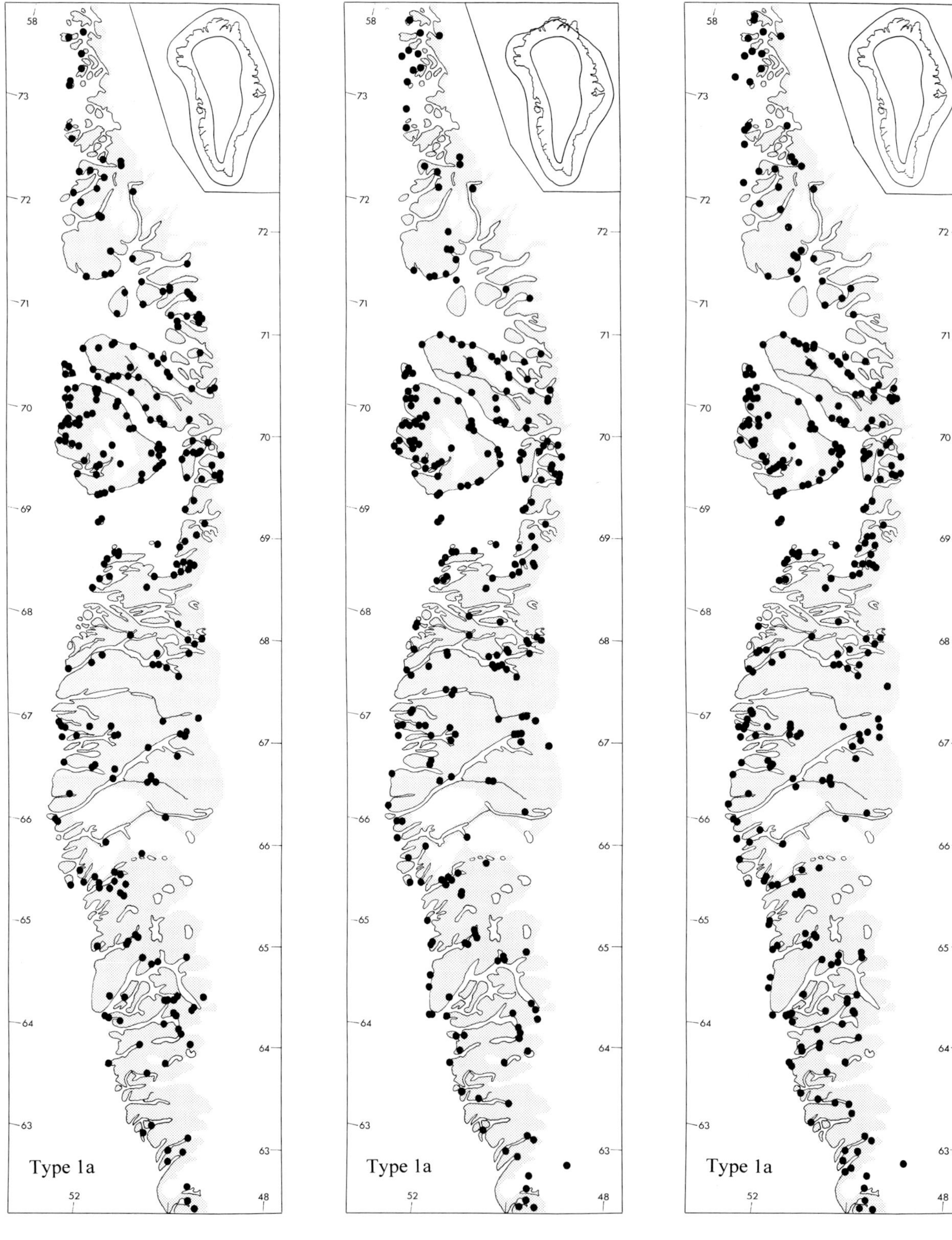

4. Dryas integrifolia

5. Festuca brachyphylla

6. Luzula confusa

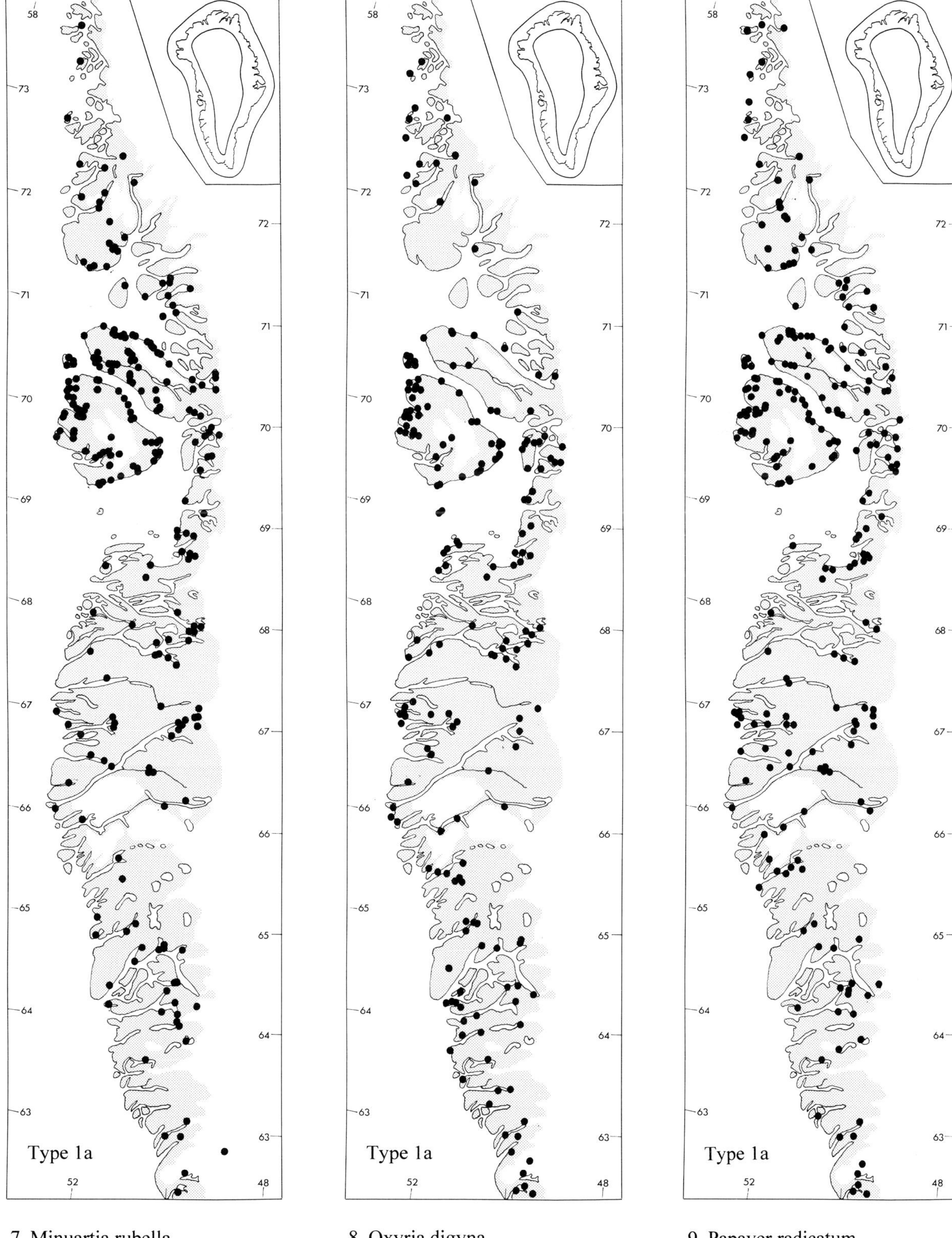

7. Minuartia rubella

8. Oxyria digyna

9. Papaver radicatum

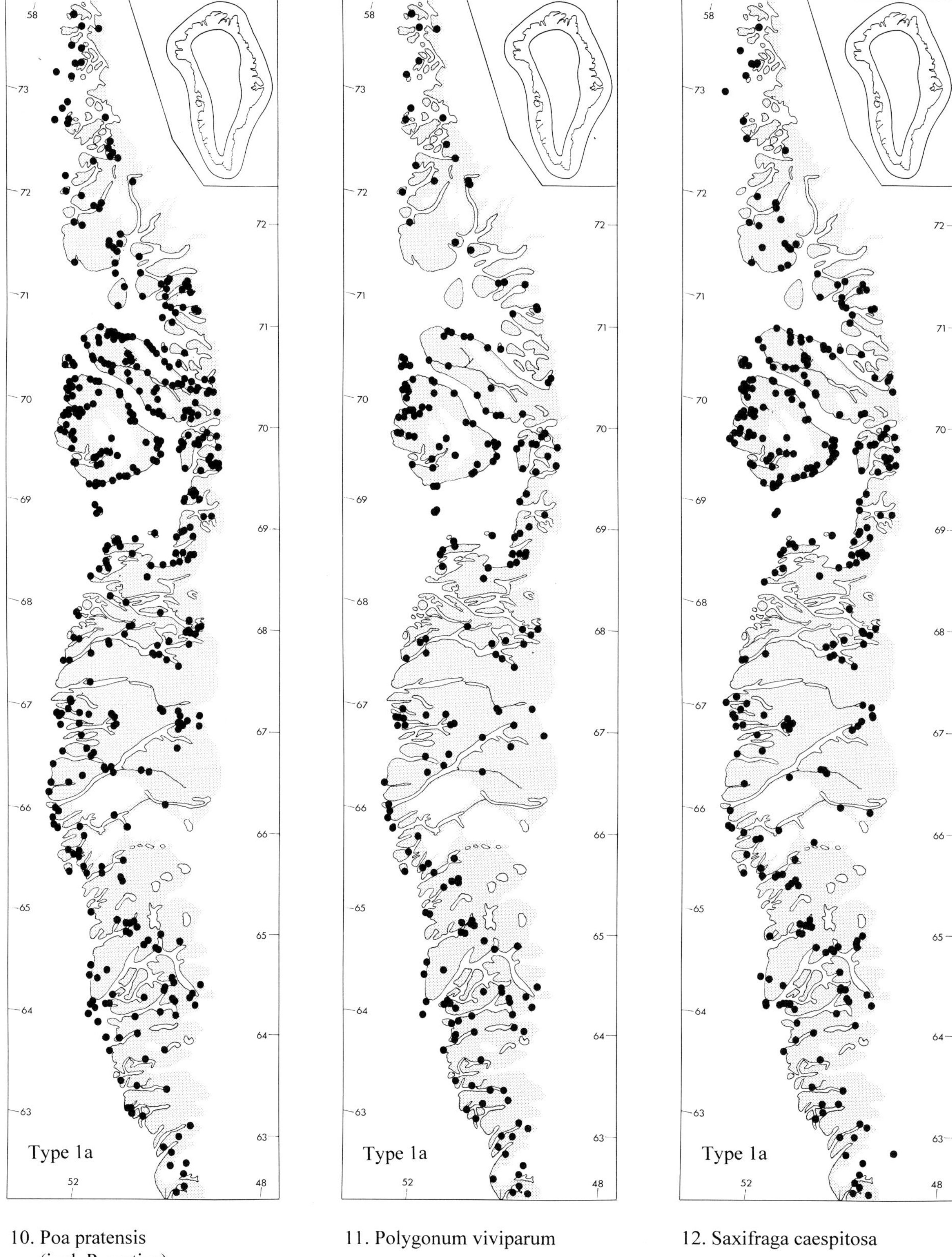

10. Poa pratensis (incl. P. arctica)

11. Polygonum viviparum

12. Saxifraga caespitosa

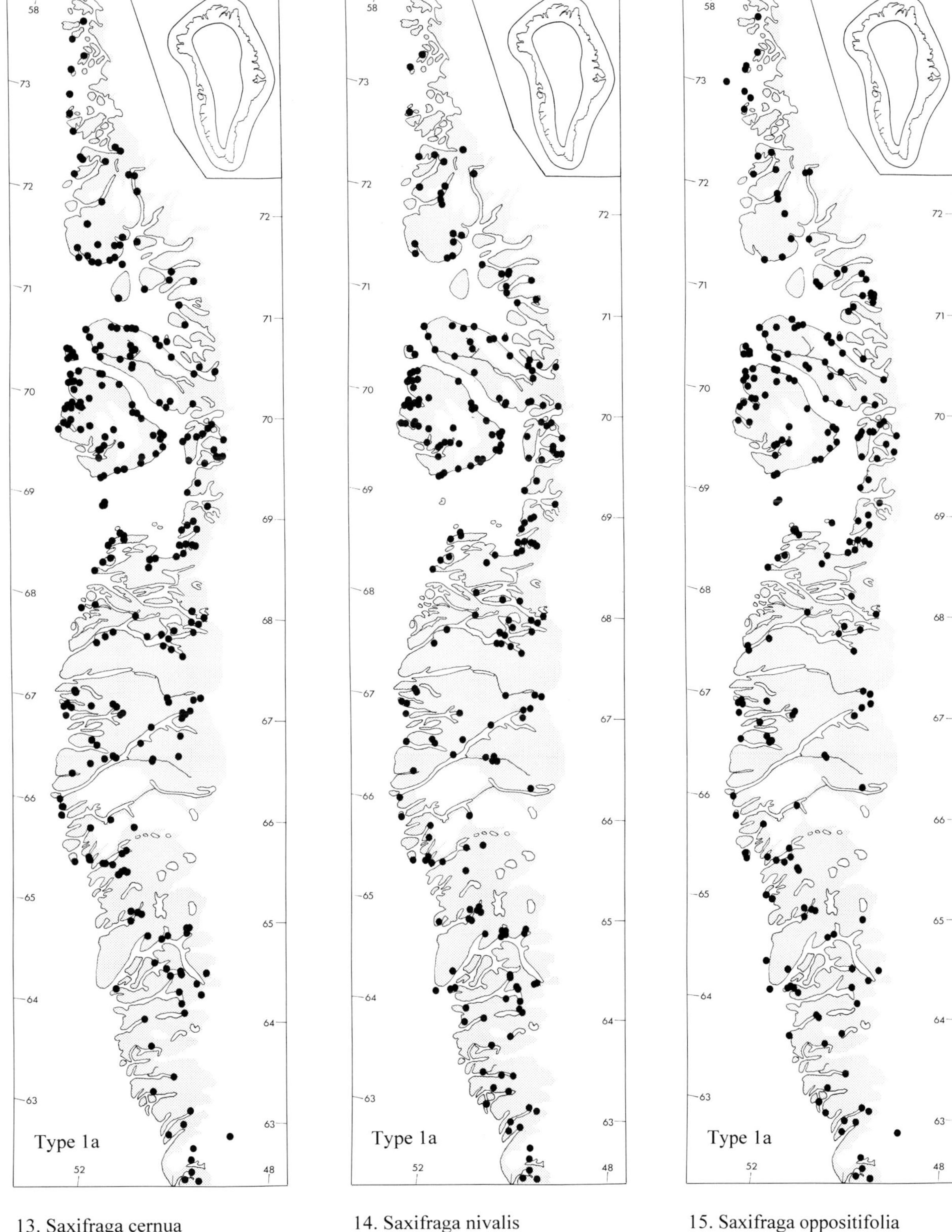

13. Saxifraga cernua

14. Saxifraga nivalis

15. Saxifraga oppositifolia

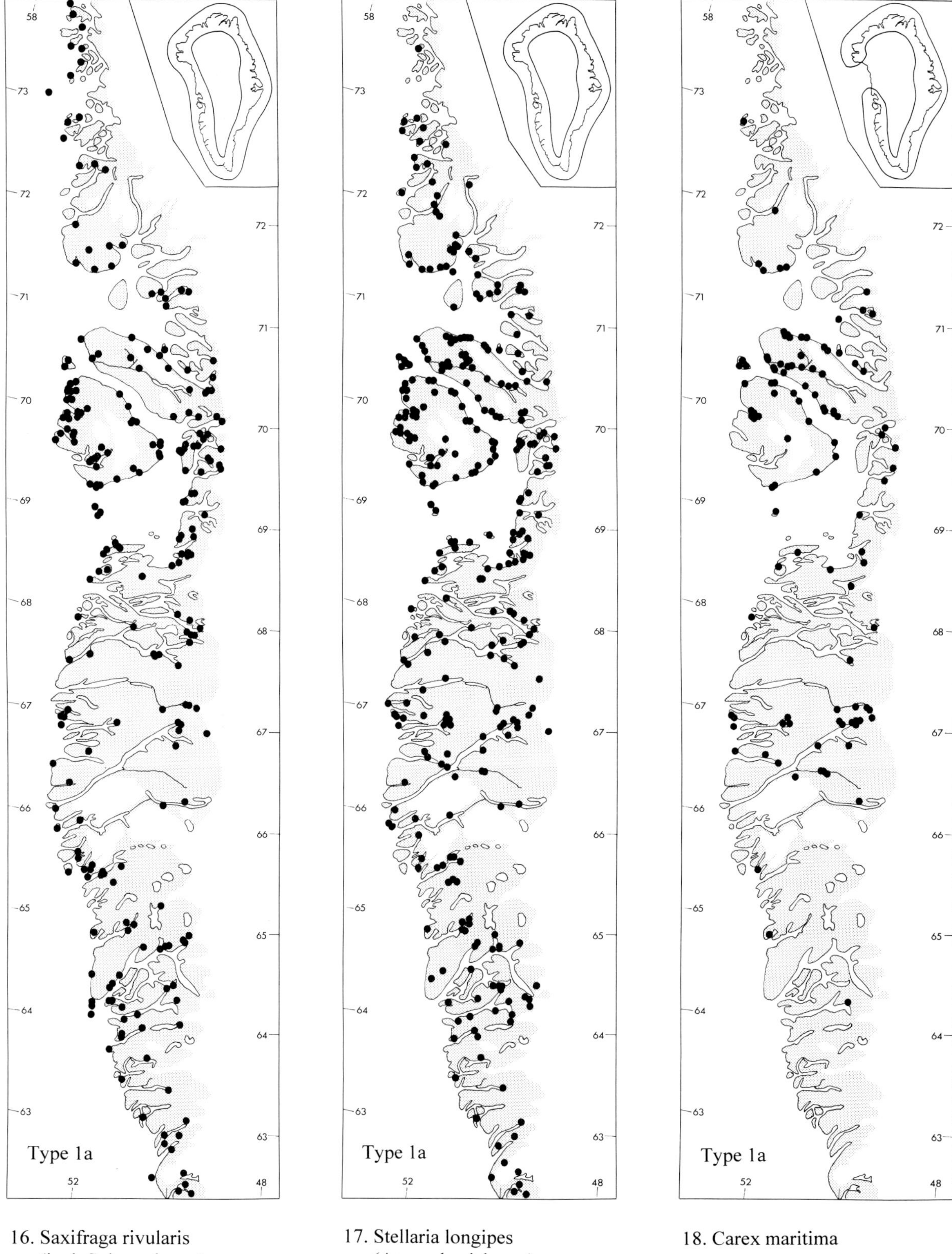

16. Saxifraga rivularis (incl. S. hyperborea)

17. Stellaria longipes (A: sepals glabrous)

18. Carex maritima

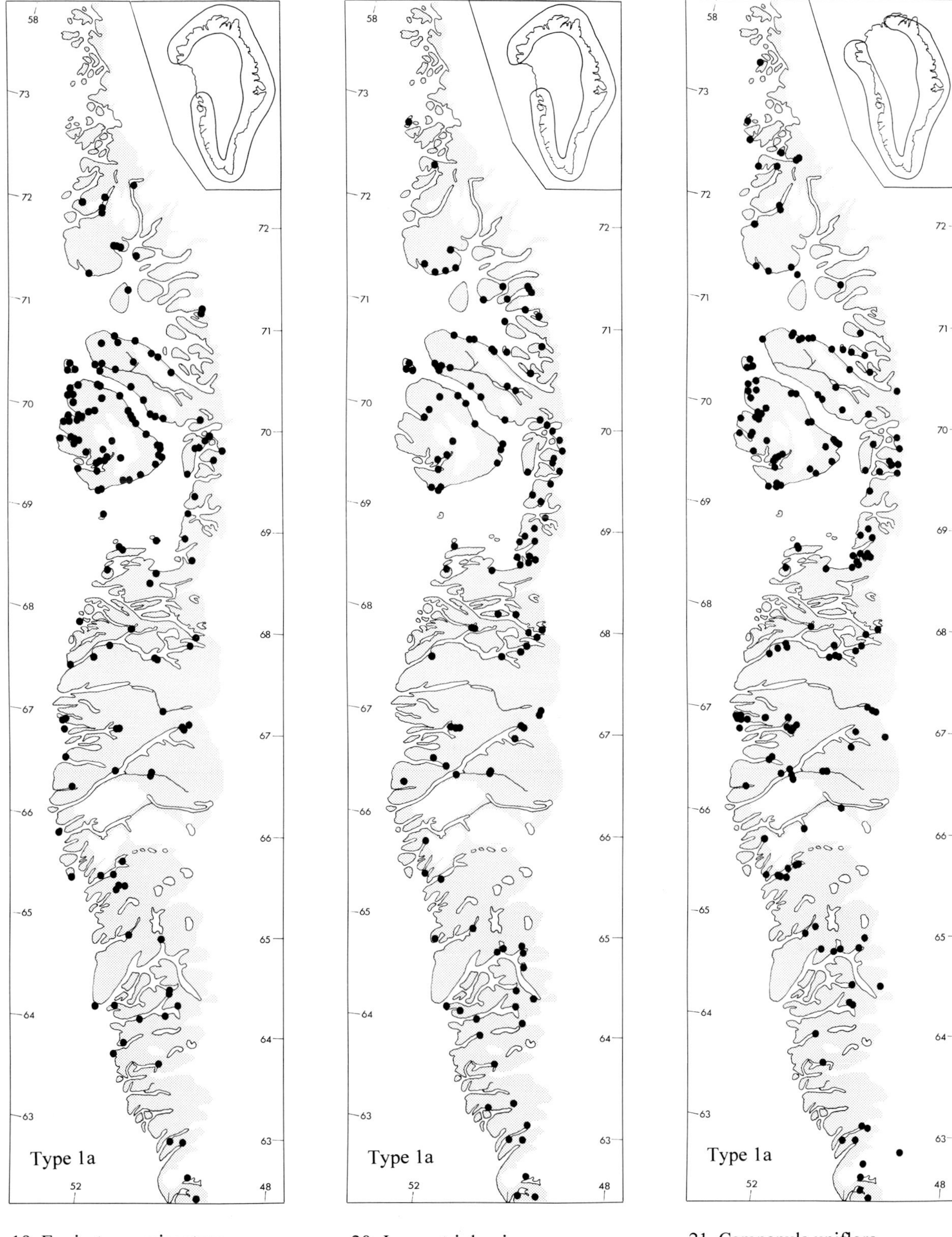

19. Equisetum variegatum

20. Juncus triglumis

21. Campanula uniflora

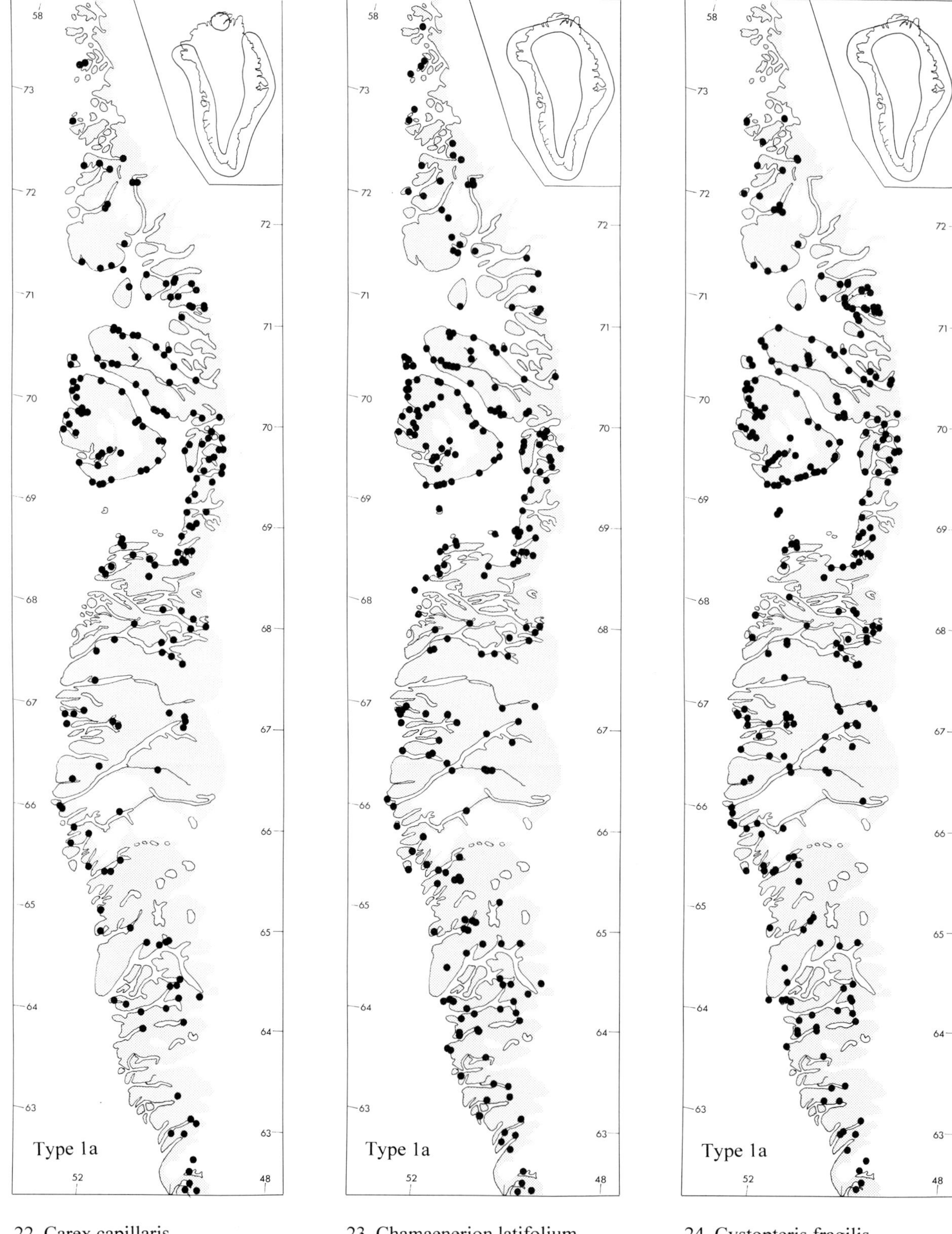

22. Carex capillaris ssp. fuscidula

23. Chamaenerion latifolium

24. Cystopteris fragilis

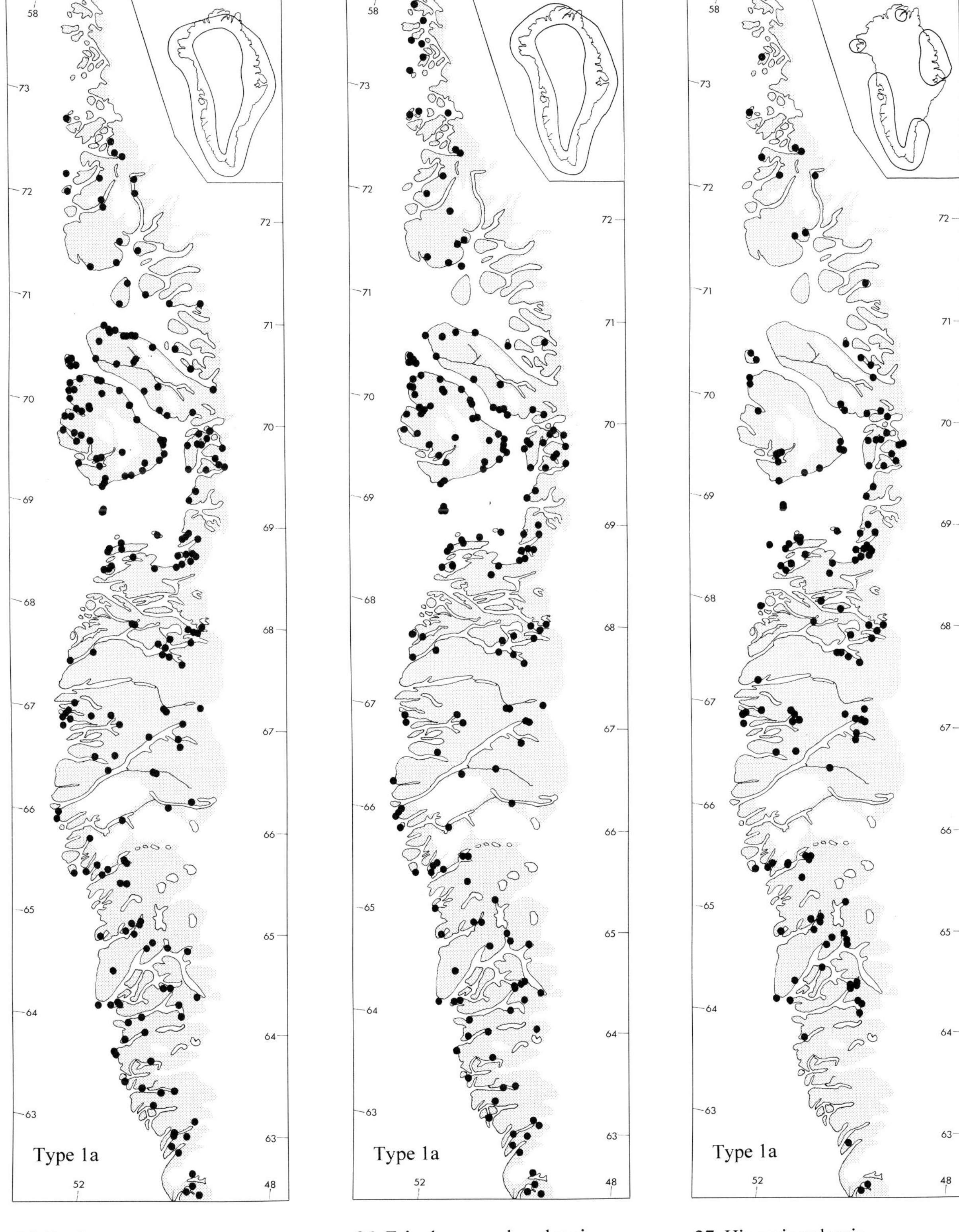

25. Equisetum arvense

26. Eriophorum scheuchzeri

27. Hippuris vulgaris

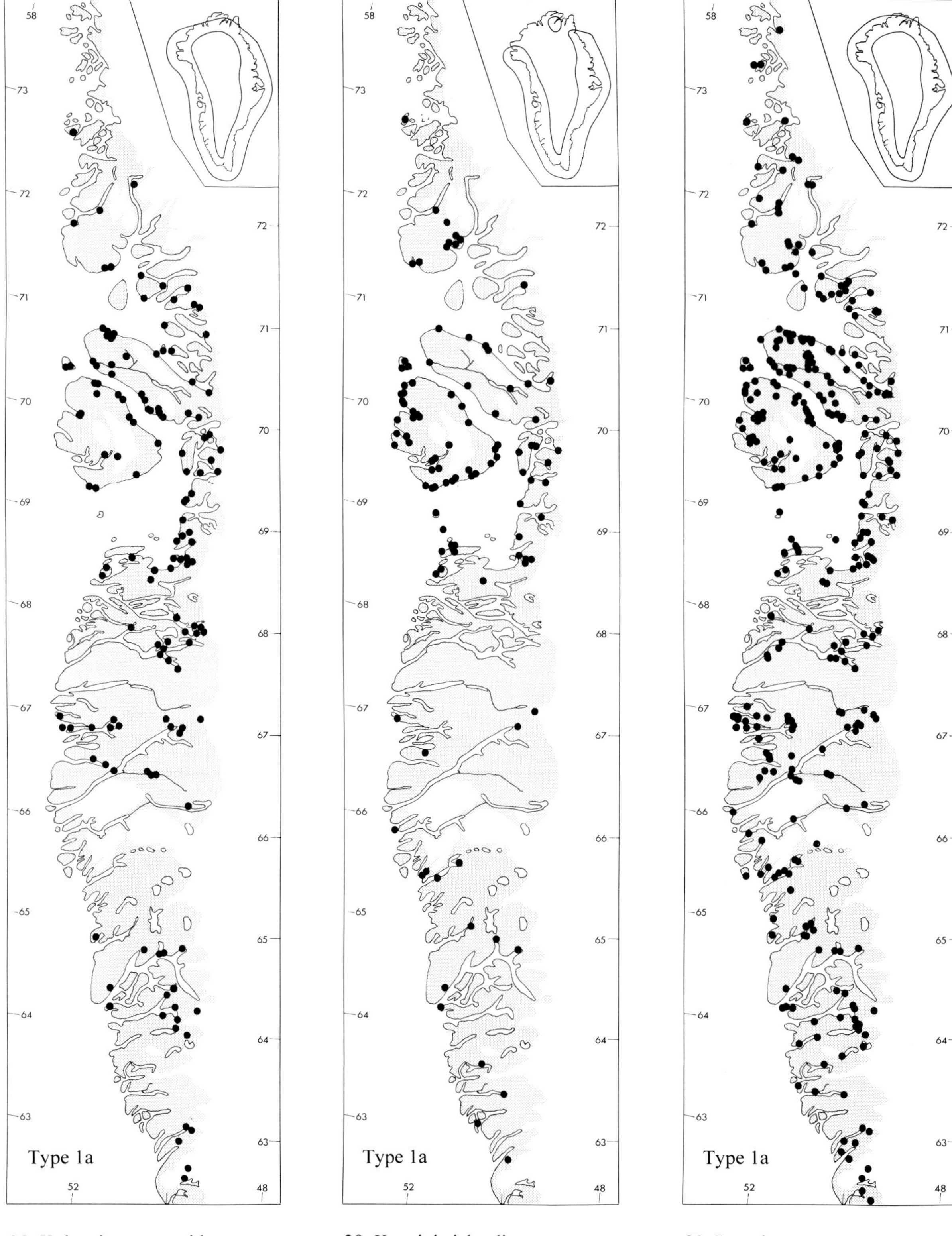

28. Kobresia myosuroides

29. Koenigia islandica

30. Poa glauca

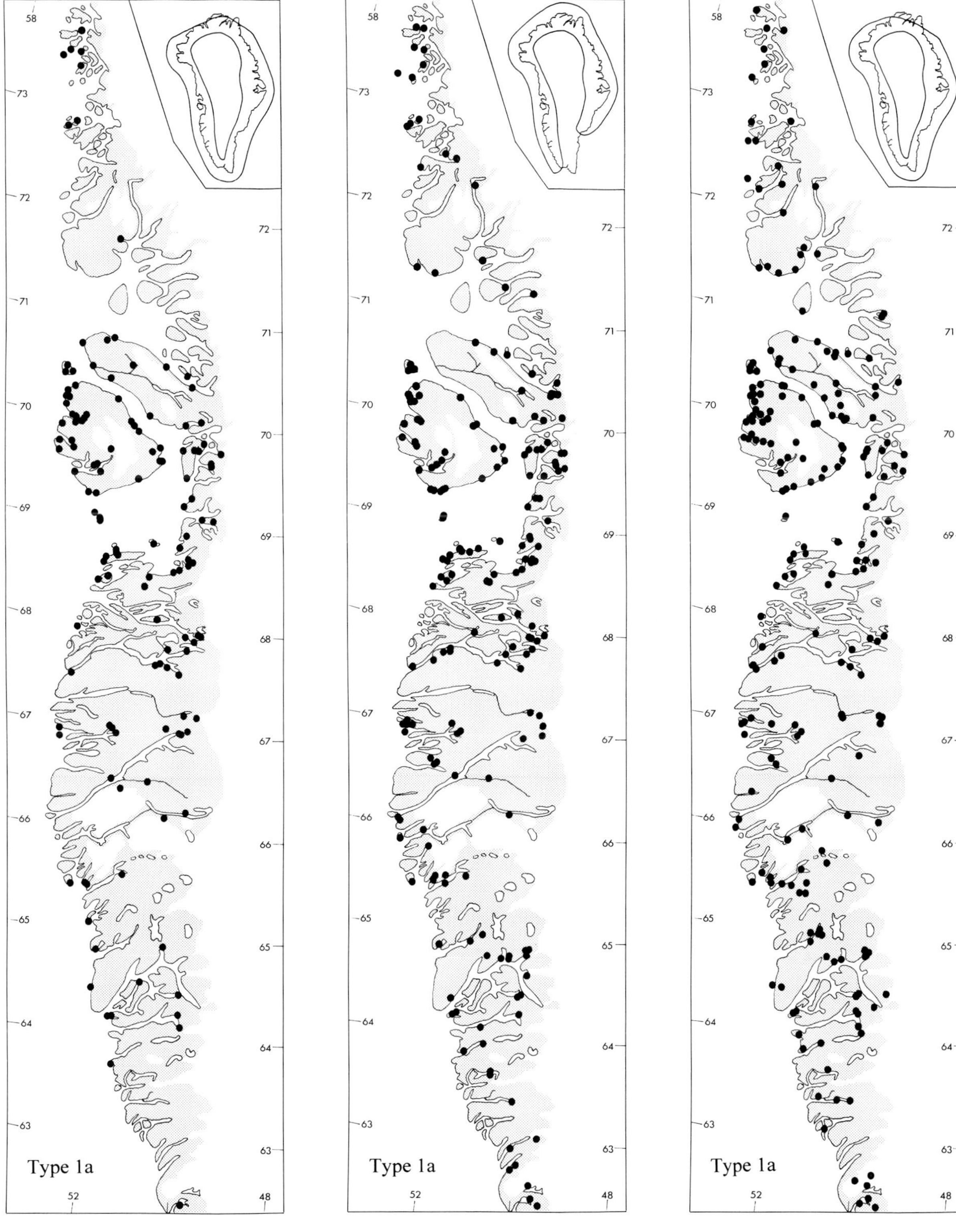

31. Ranunculus hyperboreus

32. Saxifraga foliolosa

33. Silene acaulis

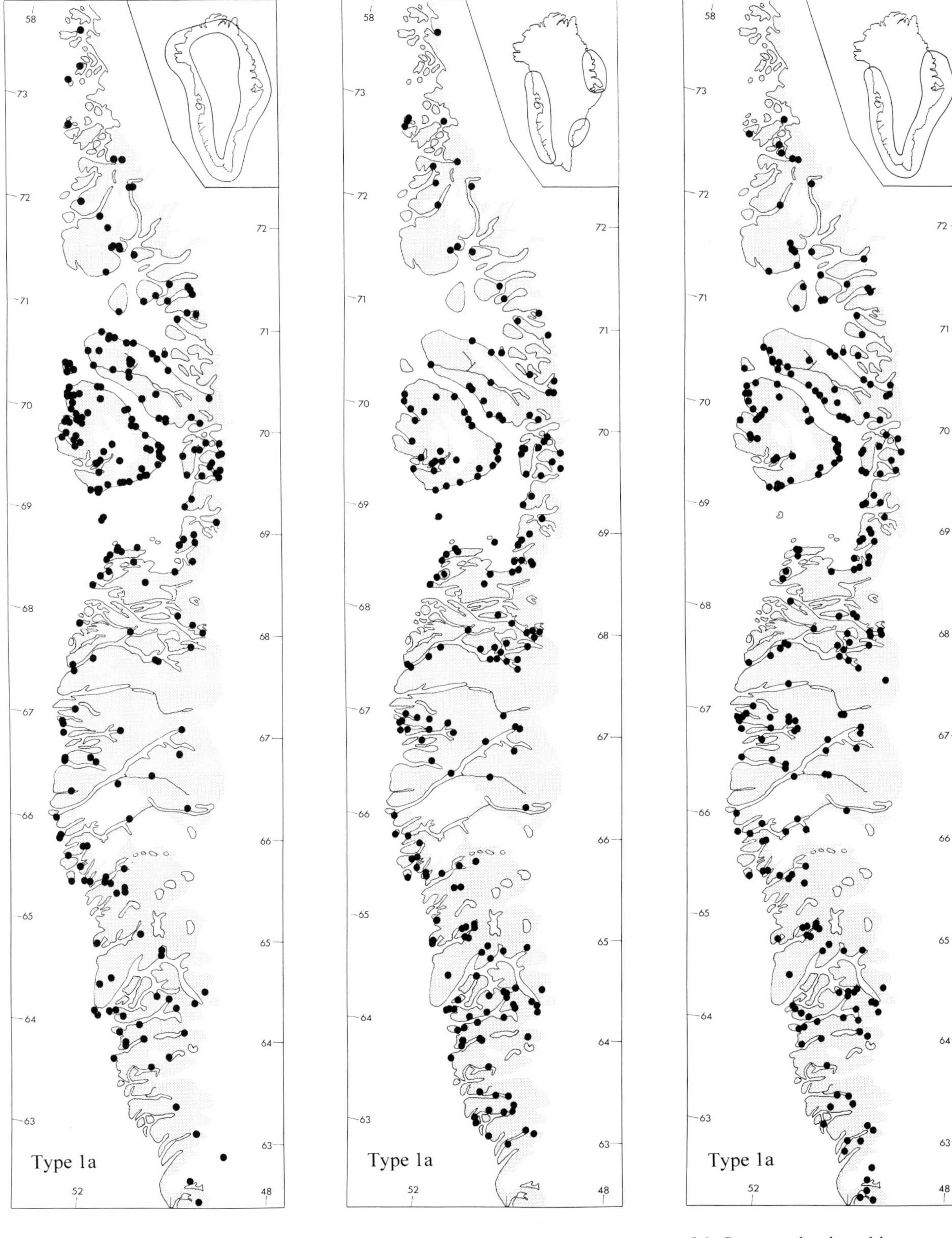

34. Trisetum spicatum

35. Betula nana

36. Campanula gieseckiana

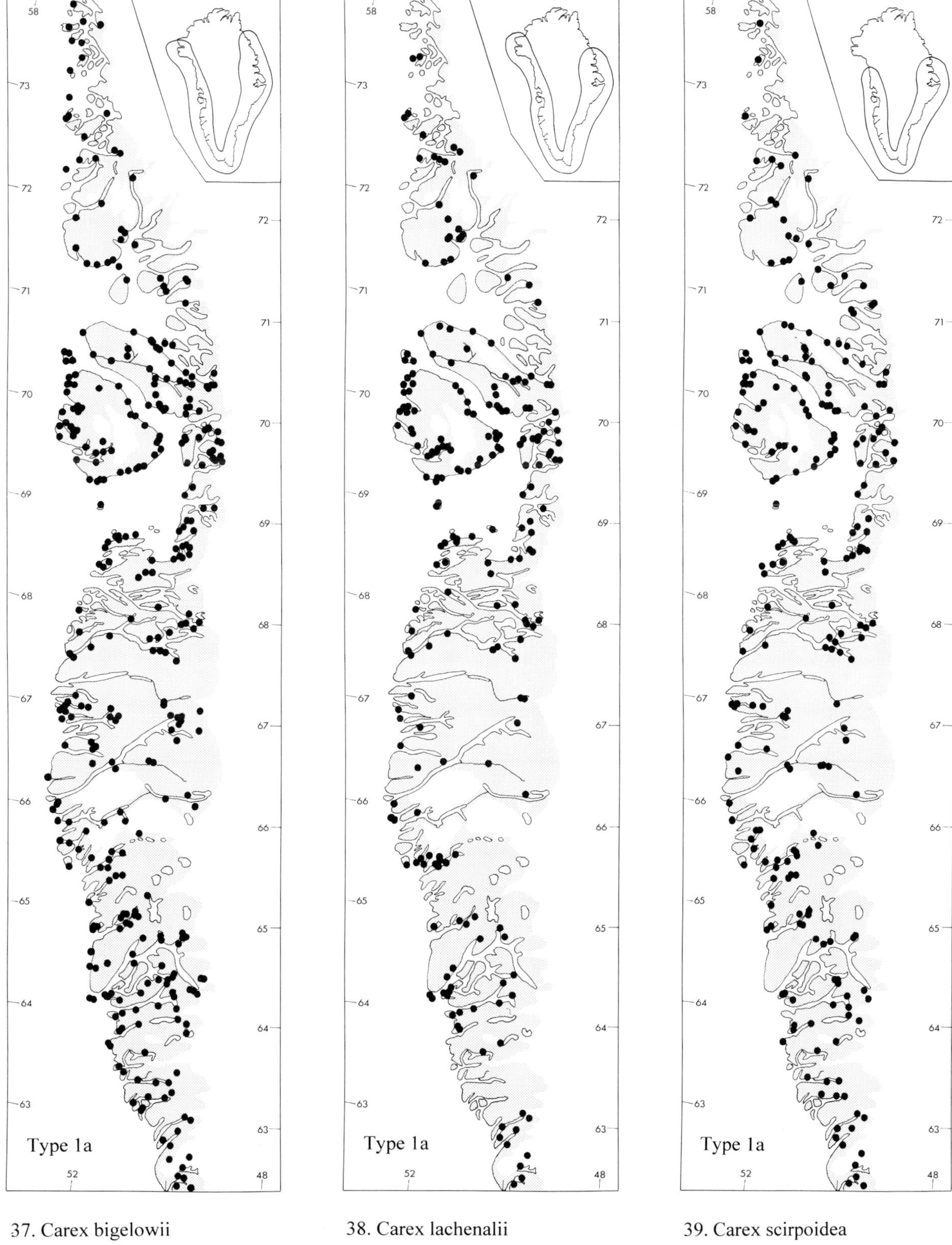

37. Carex bigelowii

38. Carex lachenalii

39. Carex scirpoidea

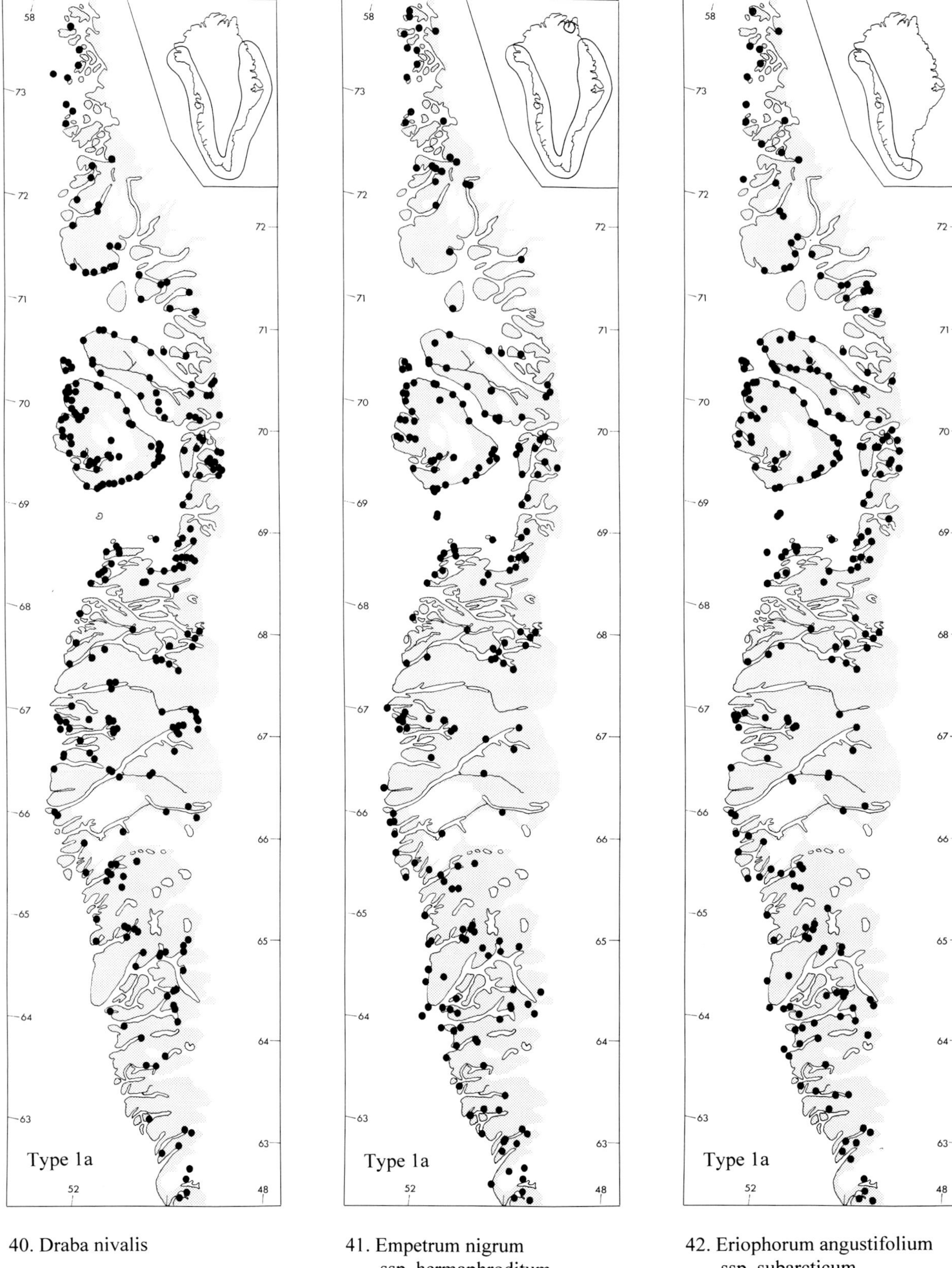

40. Draba nivalis

41. Empetrum nigrum ssp. hermaphroditum

42. Eriophorum angustifolium ssp. subarcticum

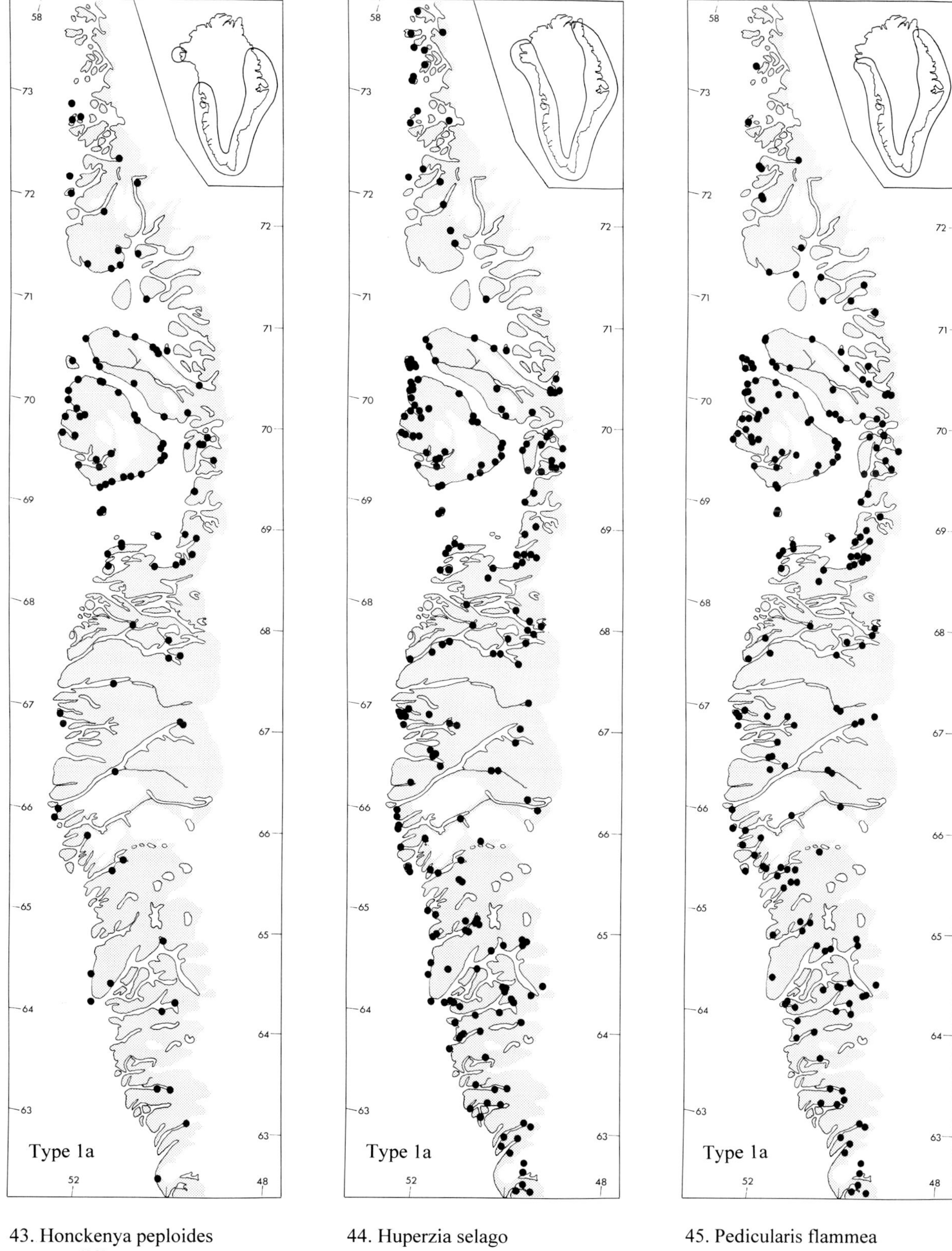

43. Honckenya peploides var. diffusa

44. Huperzia selago

45. Pedicularis flammea

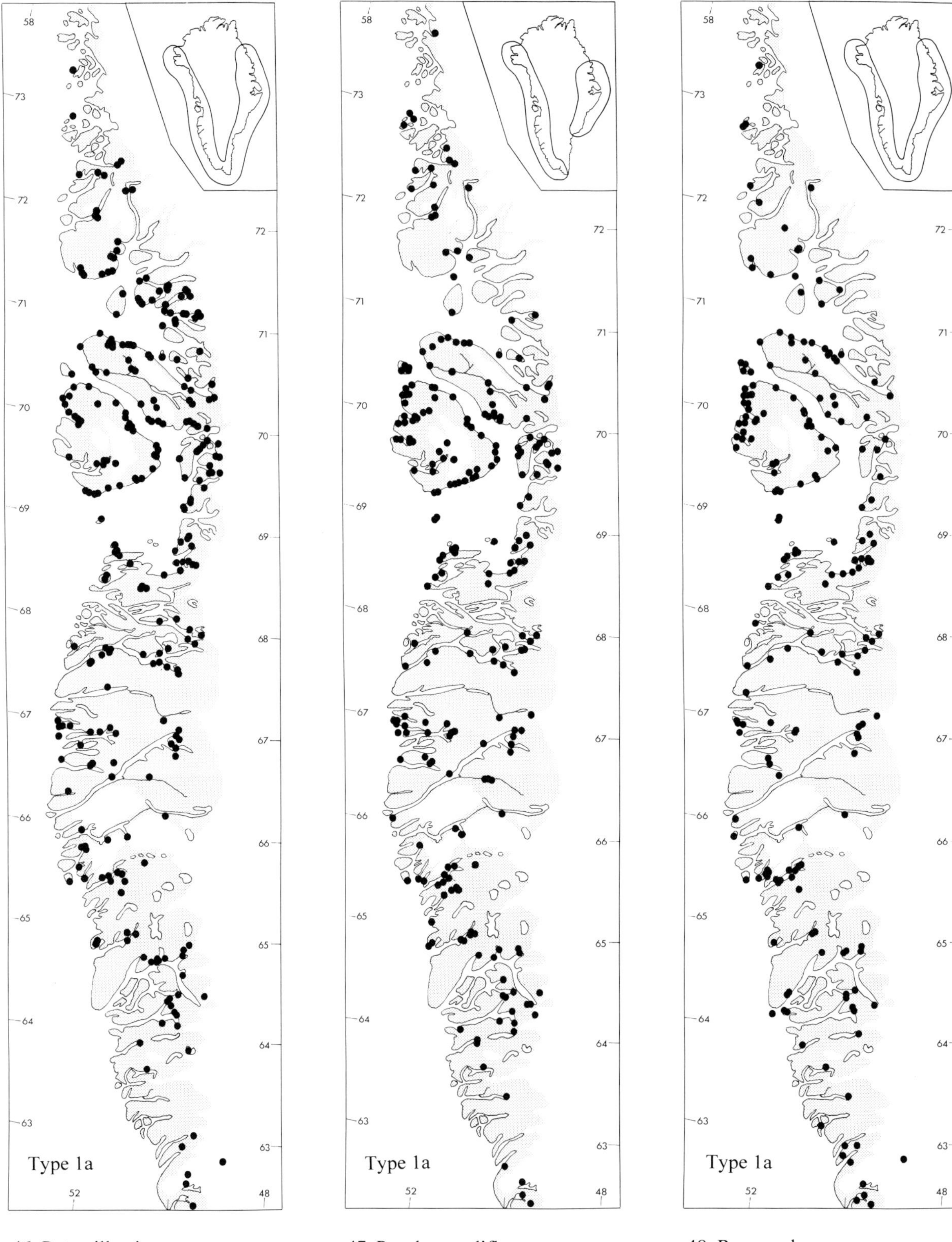

46. Potentilla nivea

47. Pyrola grandiflora

48. Ranunculus pygmaeus

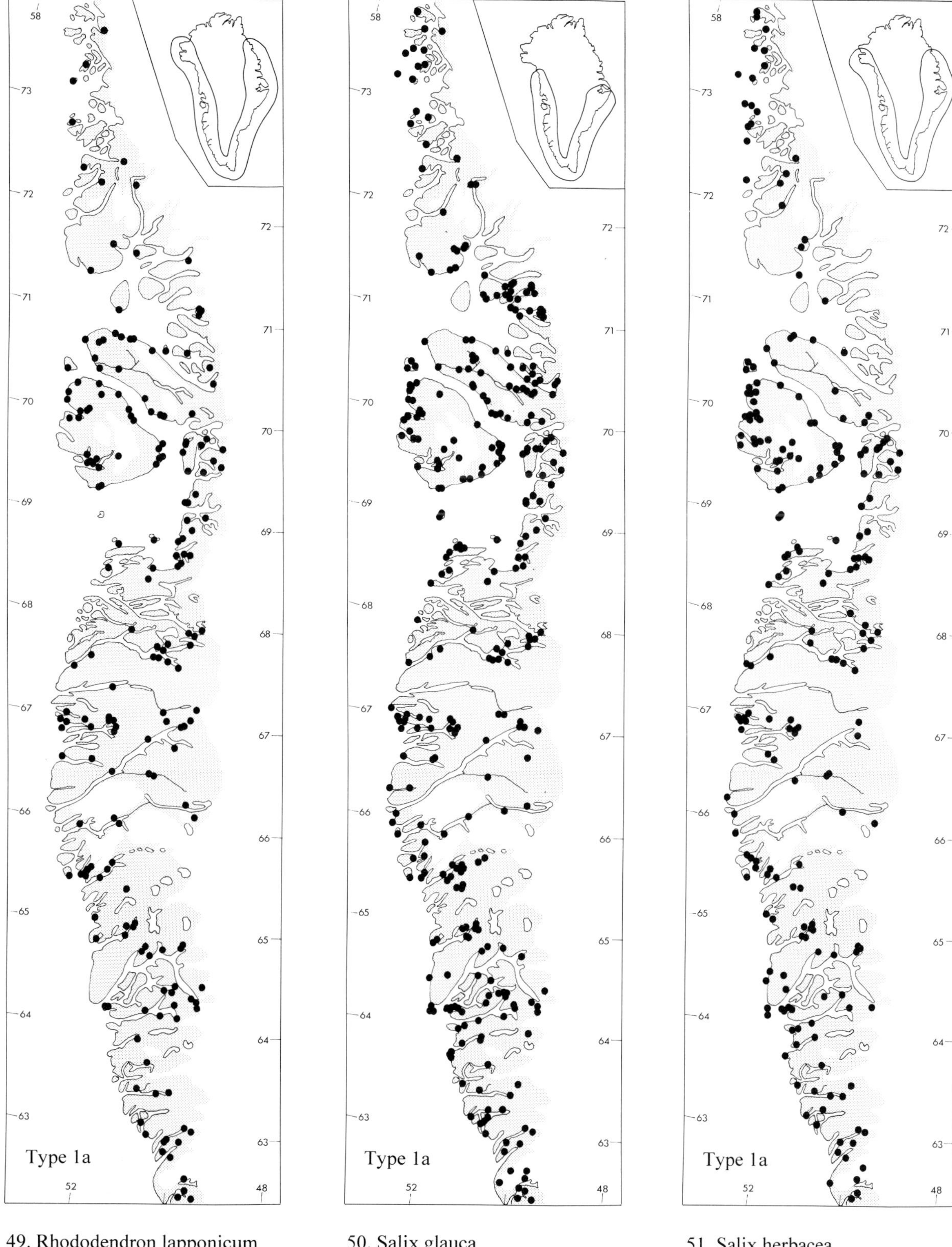

49. Rhododendron lapponicum

50. Salix glauca

51. Salix herbacea

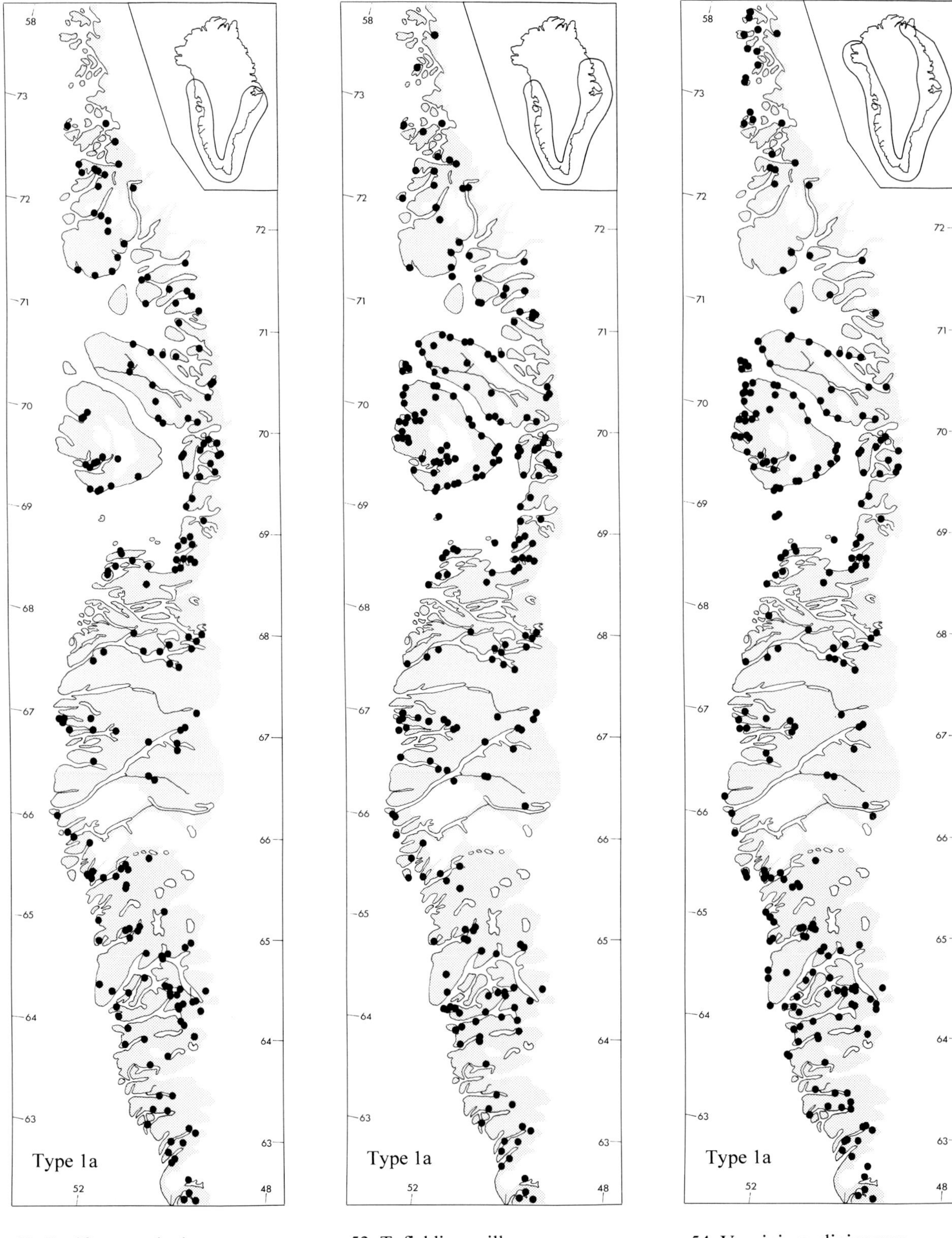

52. Saxifraga paniculata

53. Tofieldia pusilla

54. Vaccinium uliginosum

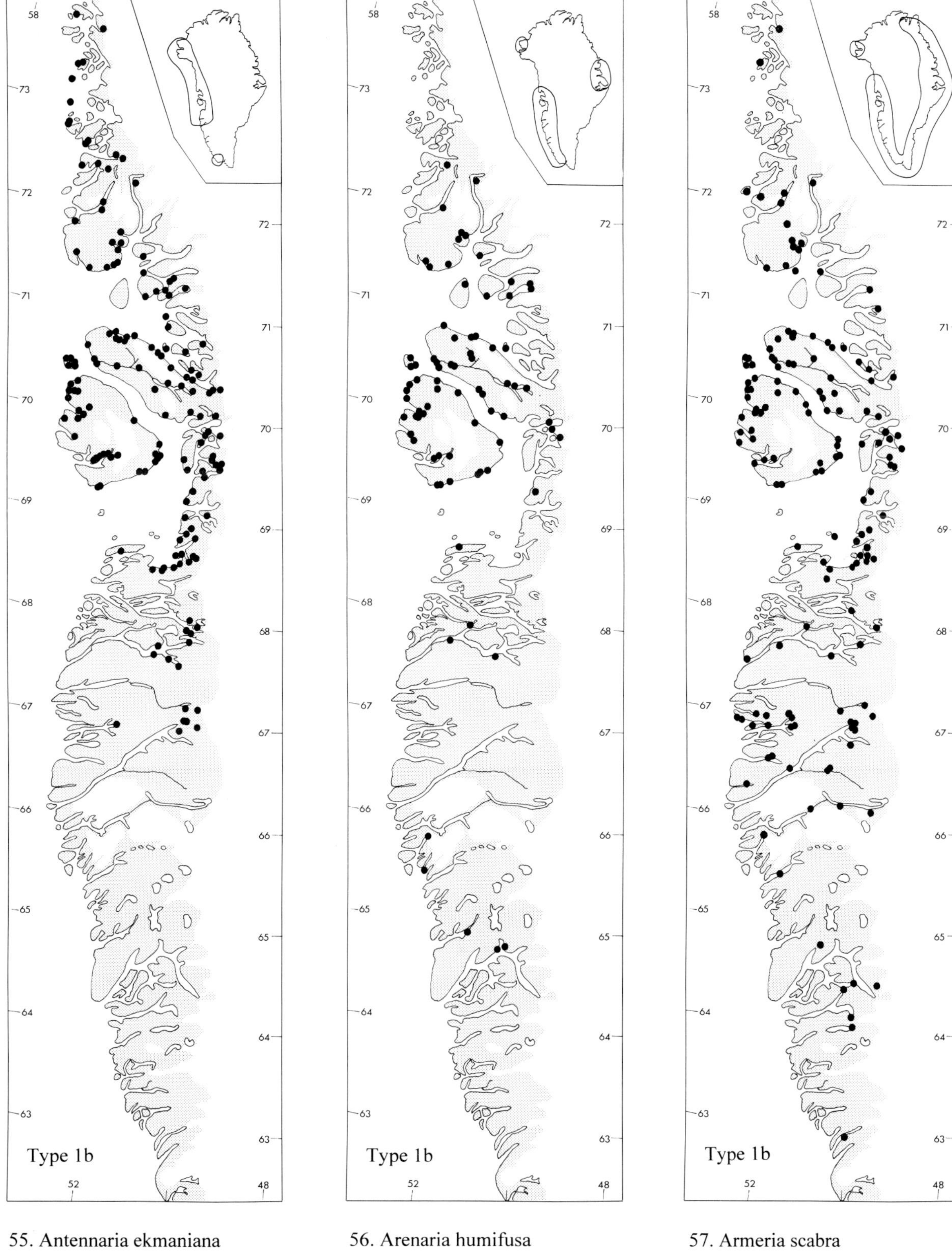

55. Antennaria ekmaniana

56. Arenaria humifusa

57. Armeria scabra
ssp. sibirica

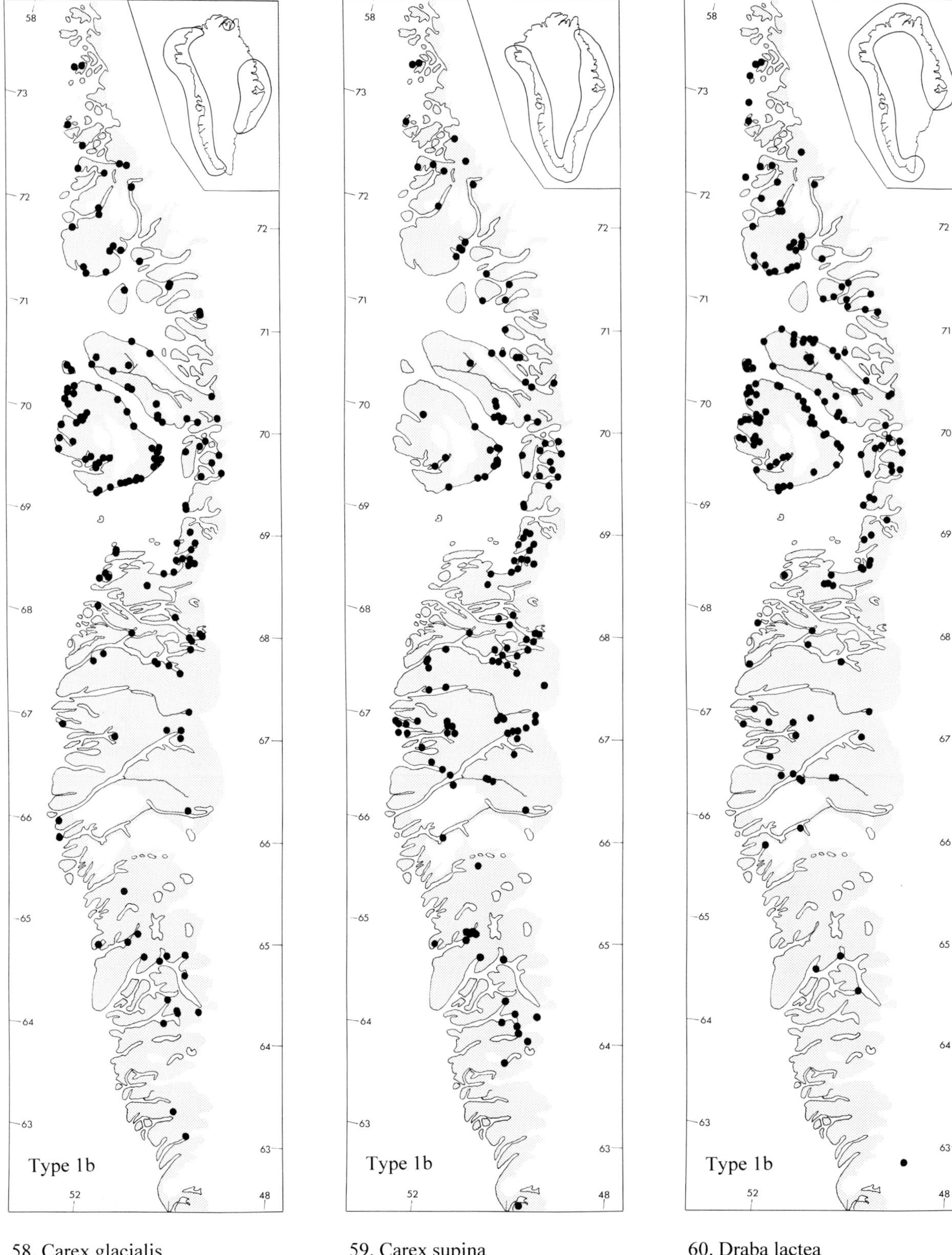

58. Carex glacialis

59. Carex supina
ssp. spaniocarpa

60. Draba lactea

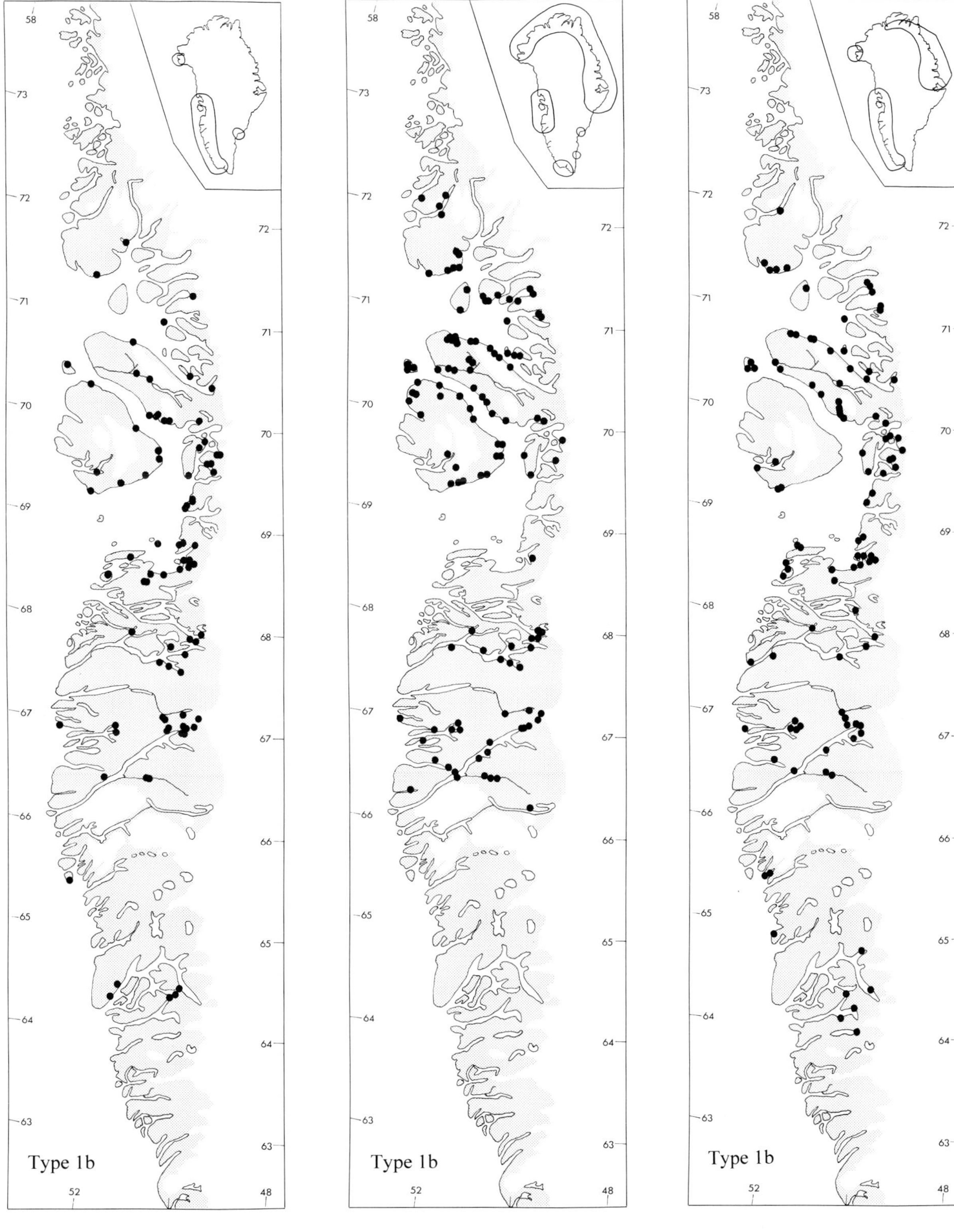

61. Eleocharis acicularis

62. Erigeron compositus

63. Kobresia simpliciuscula

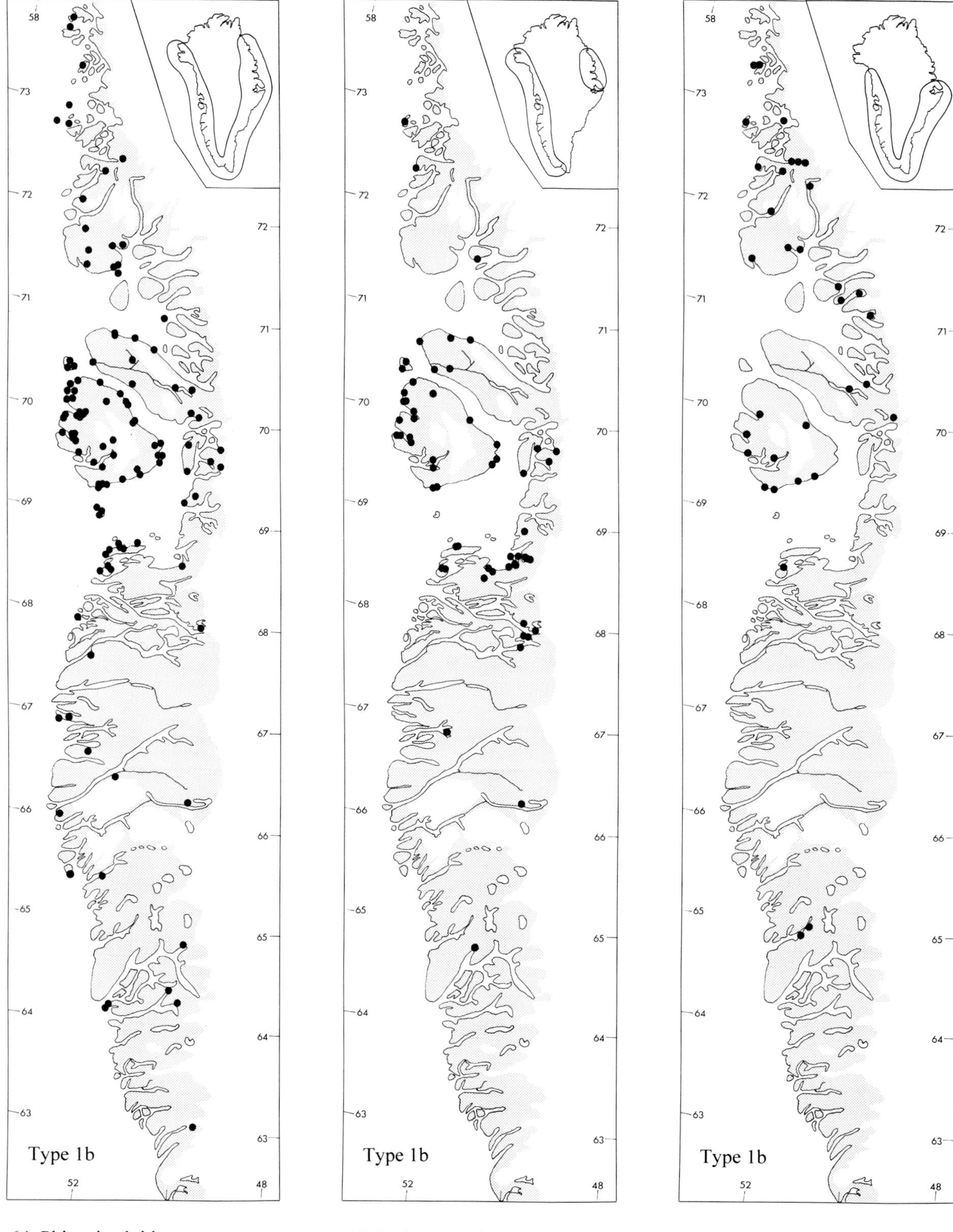

64. Phippsia algida

65. Sagina caespitosa

66. Woodsia alpina

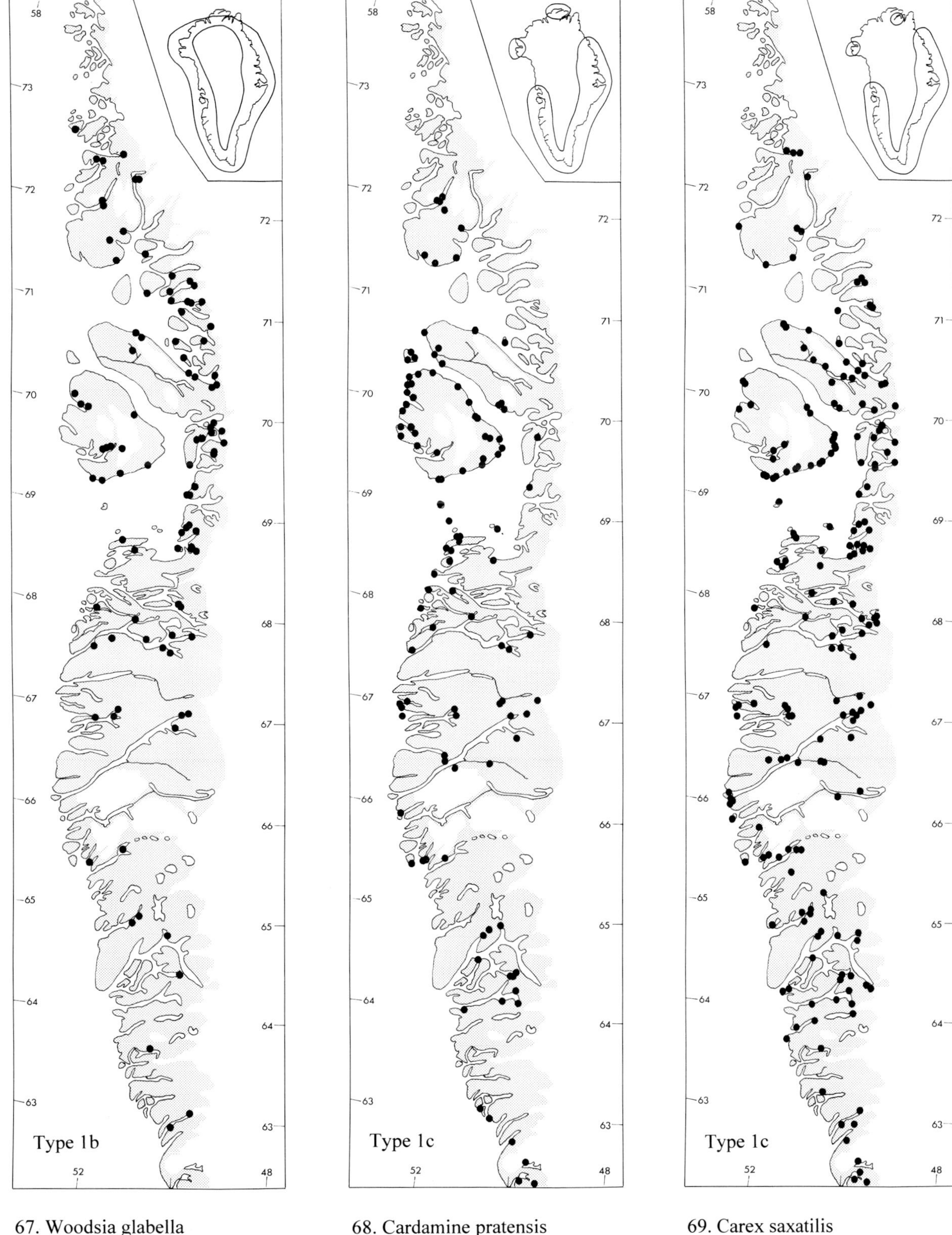

67. Woodsia glabella

68. Cardamine pratensis

69. Carex saxatilis

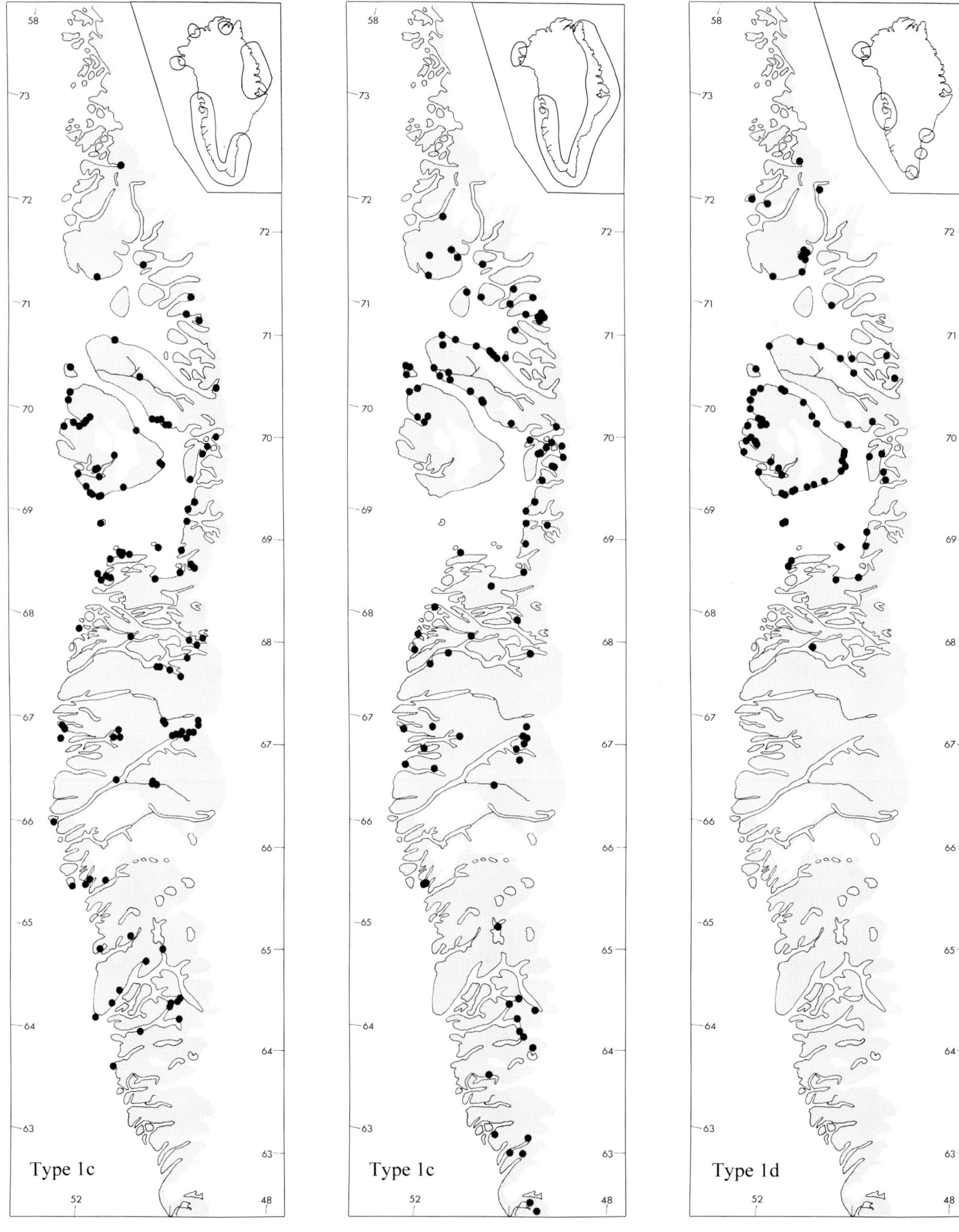

70. Ranunculus confervoides

71. Saxifraga aizoides

72. Mertensia maritima

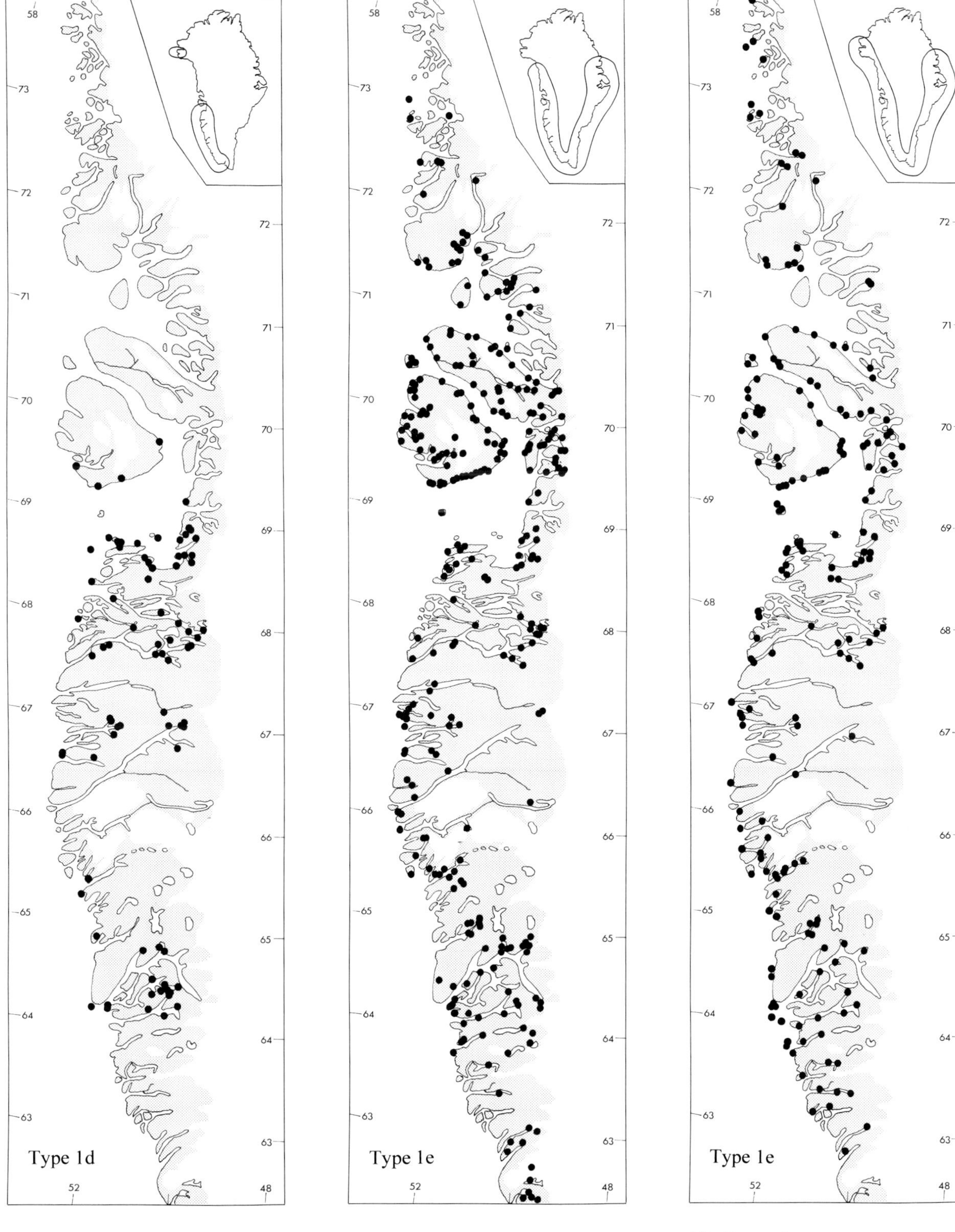

73. Vaccinium vitis-idaea ssp. minus

74. Antennaria canescens

75. Carex glareosa

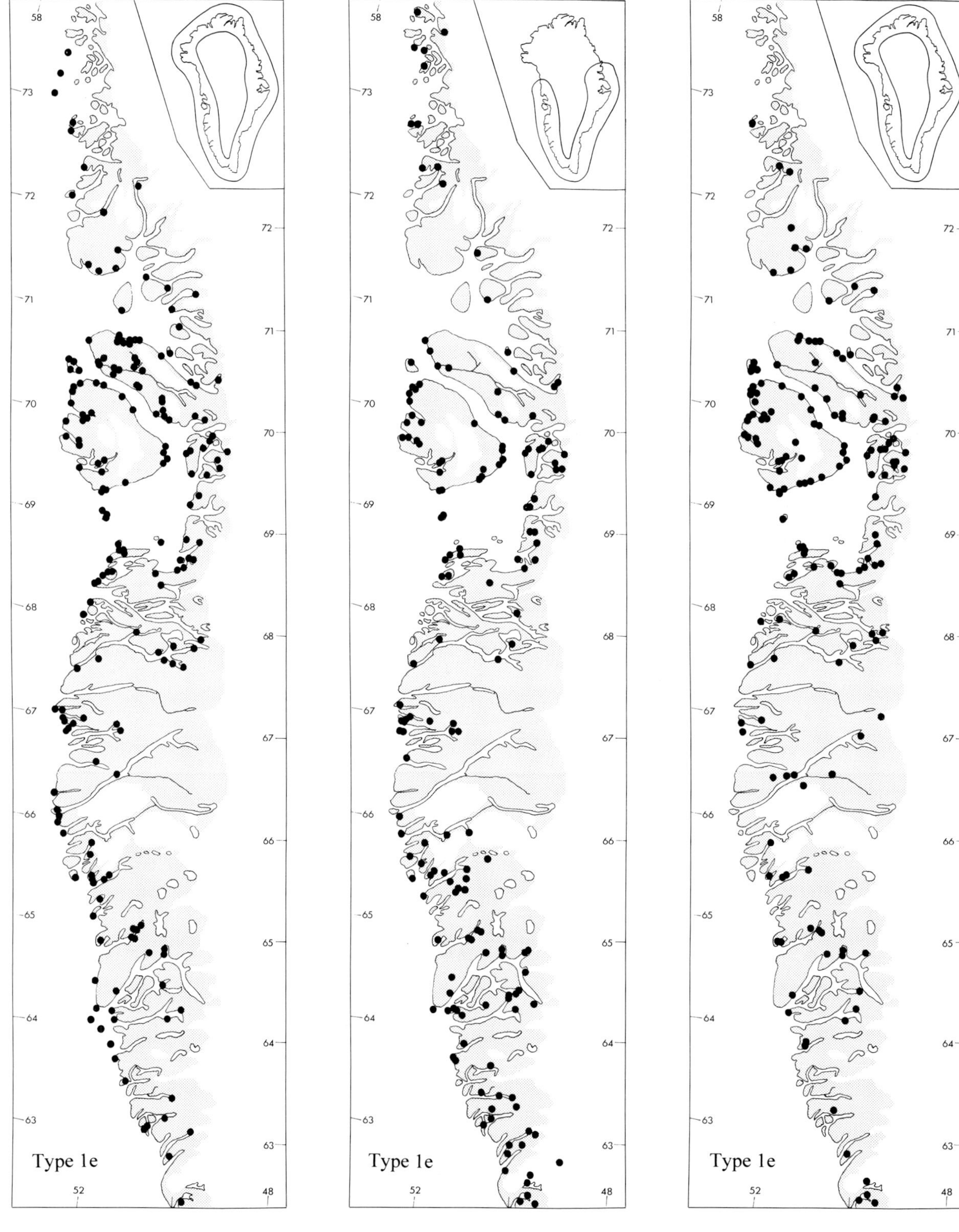

76. Cochlearia groenlandica

77. Harrimanella hypnoides

78. Juncus biglumis

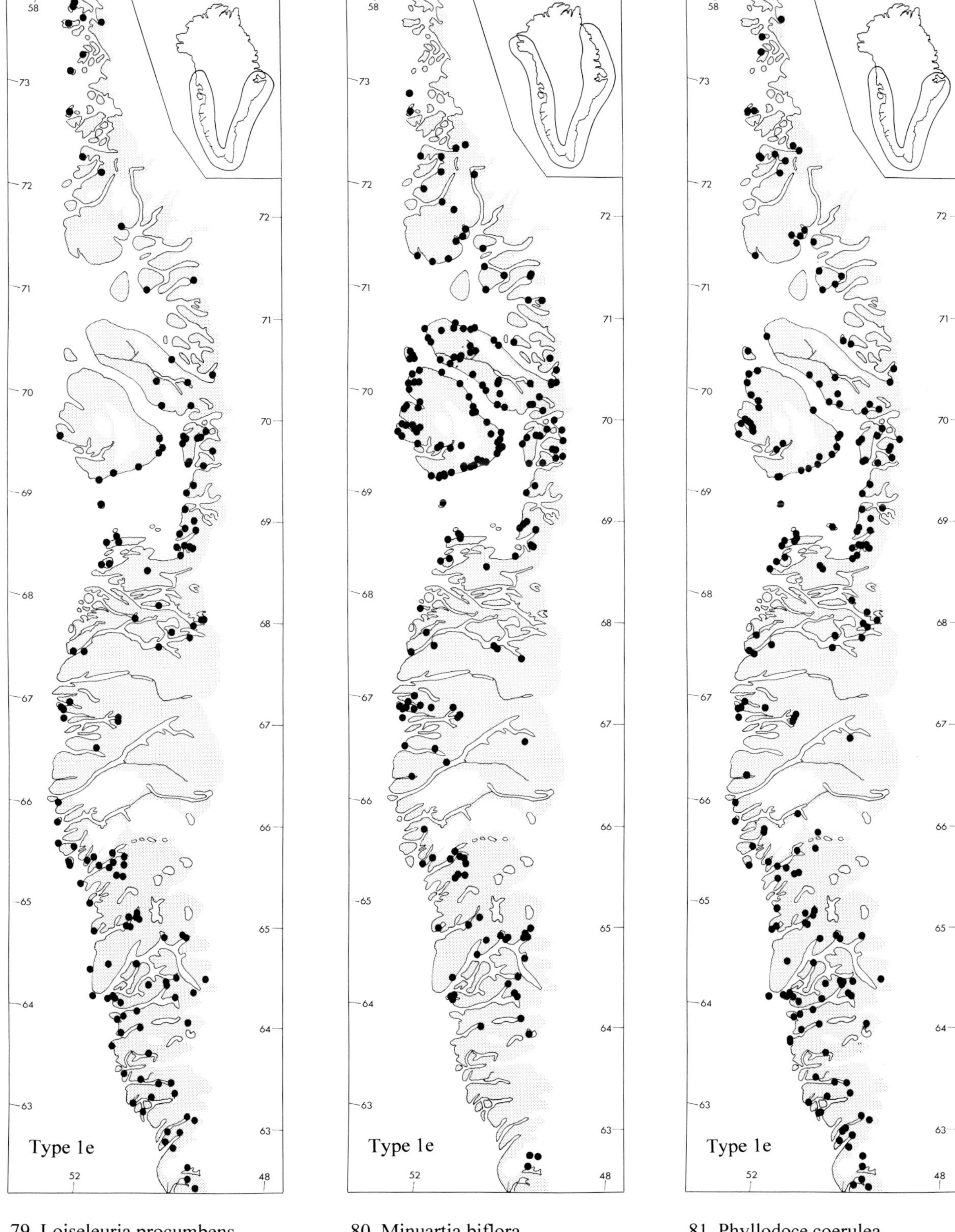

79. Loiseleuria procumbens

80. Minuartia biflora

81. Phyllodoce coerulea

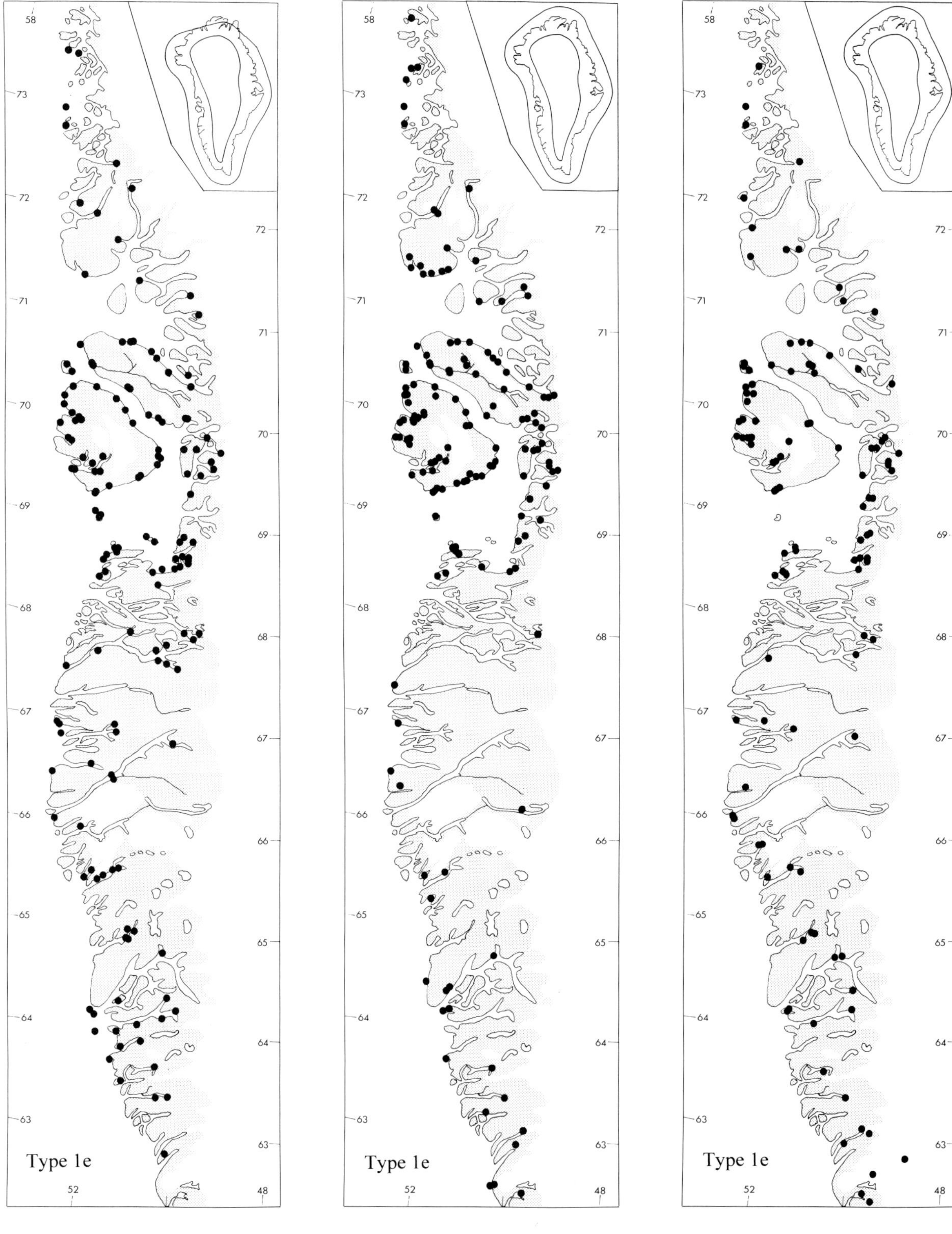

82. Puccinellia phryganodes

83. Sagina intermedia

84. Saxifraga tenuis

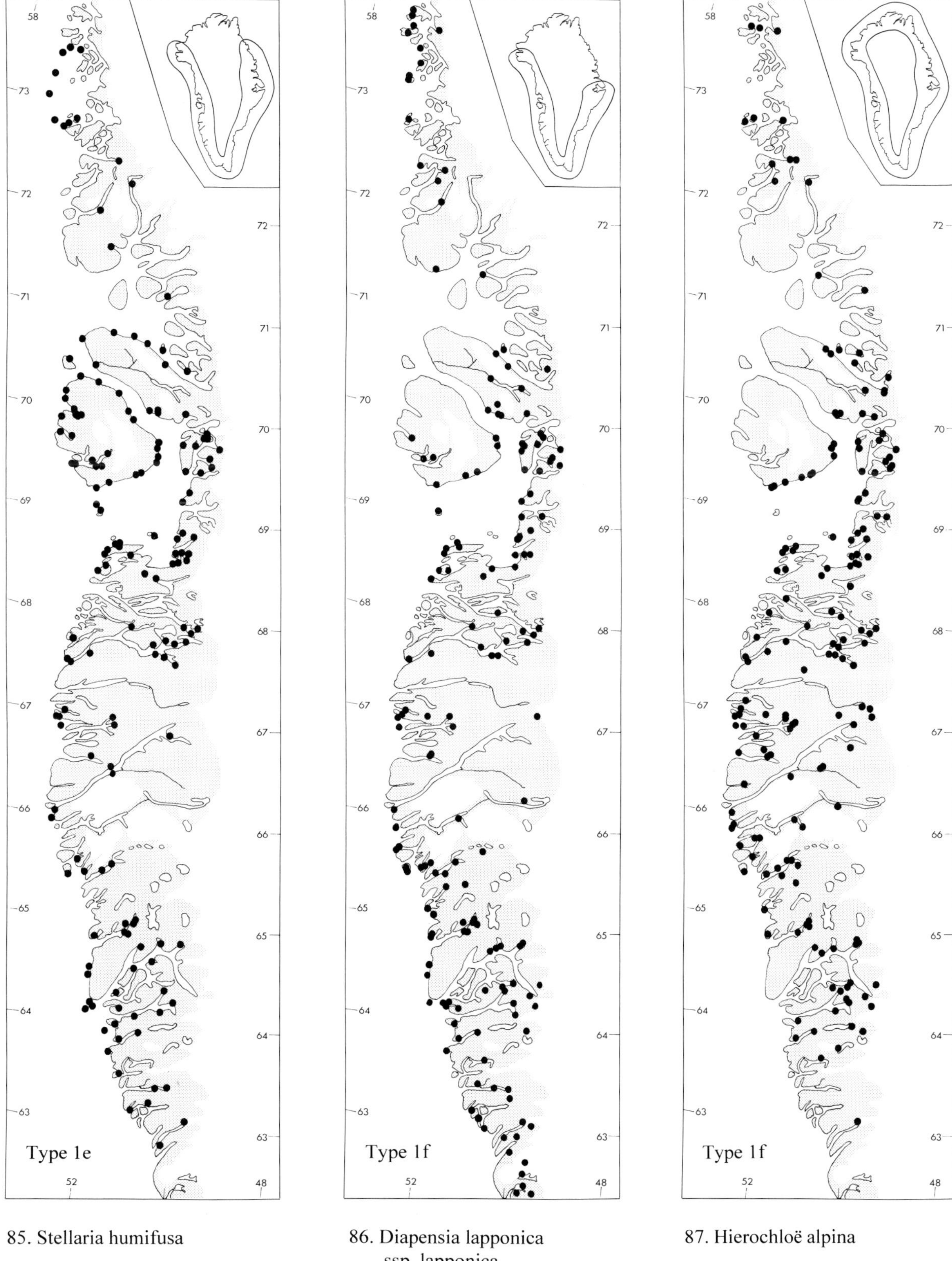

85. Stellaria humifusa

86. Diapensia lapponica ssp. lapponica

87. Hierochloë alpina

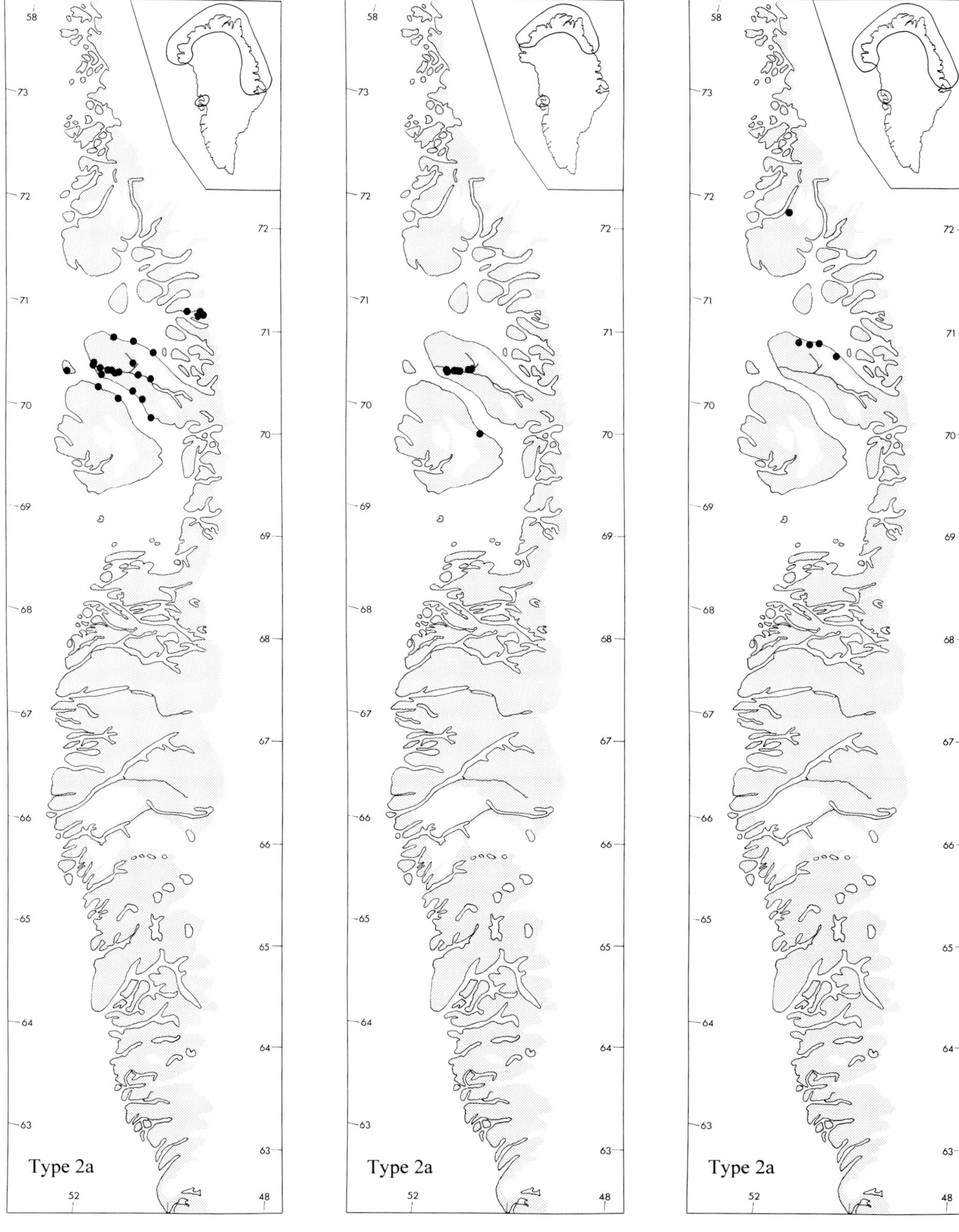

88. Braya purpurascens

89. Braya thorild-wulffii

90. Draba adamsii

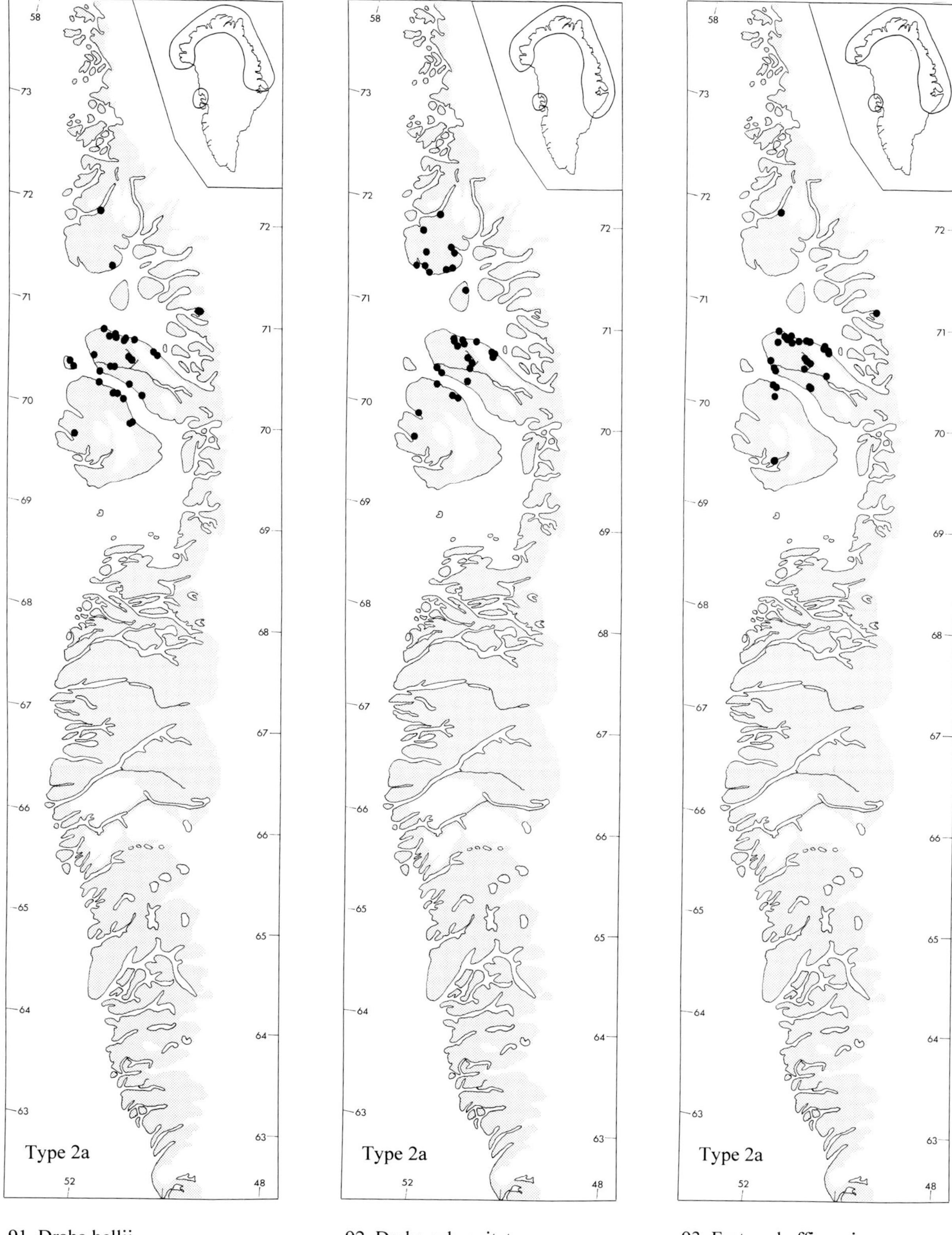

91. Draba bellii

92. Draba subcapitata

93. Festuca baffinensis

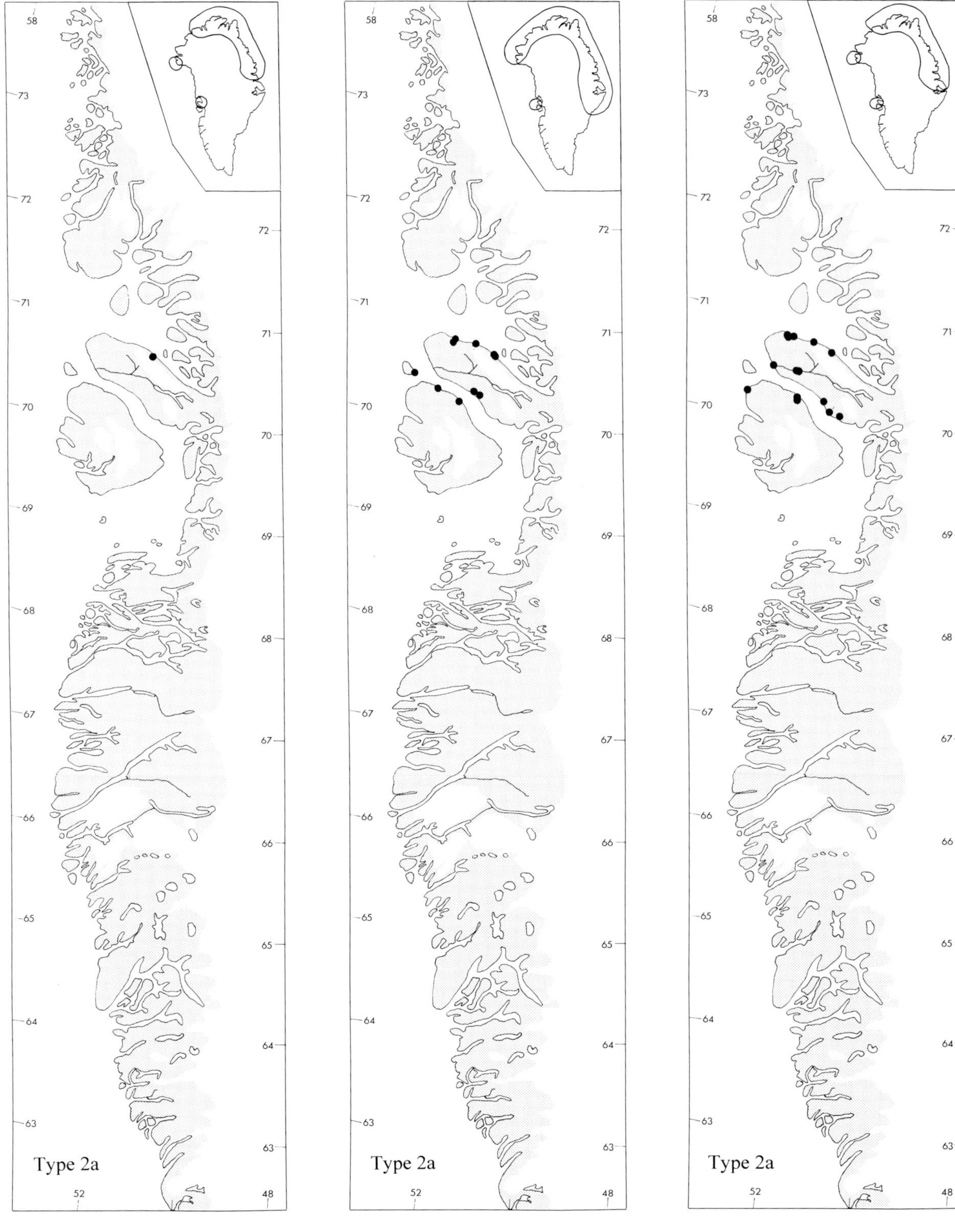

94. Minuartia rossii

95. Poa abbreviata

96. Poa hartzii

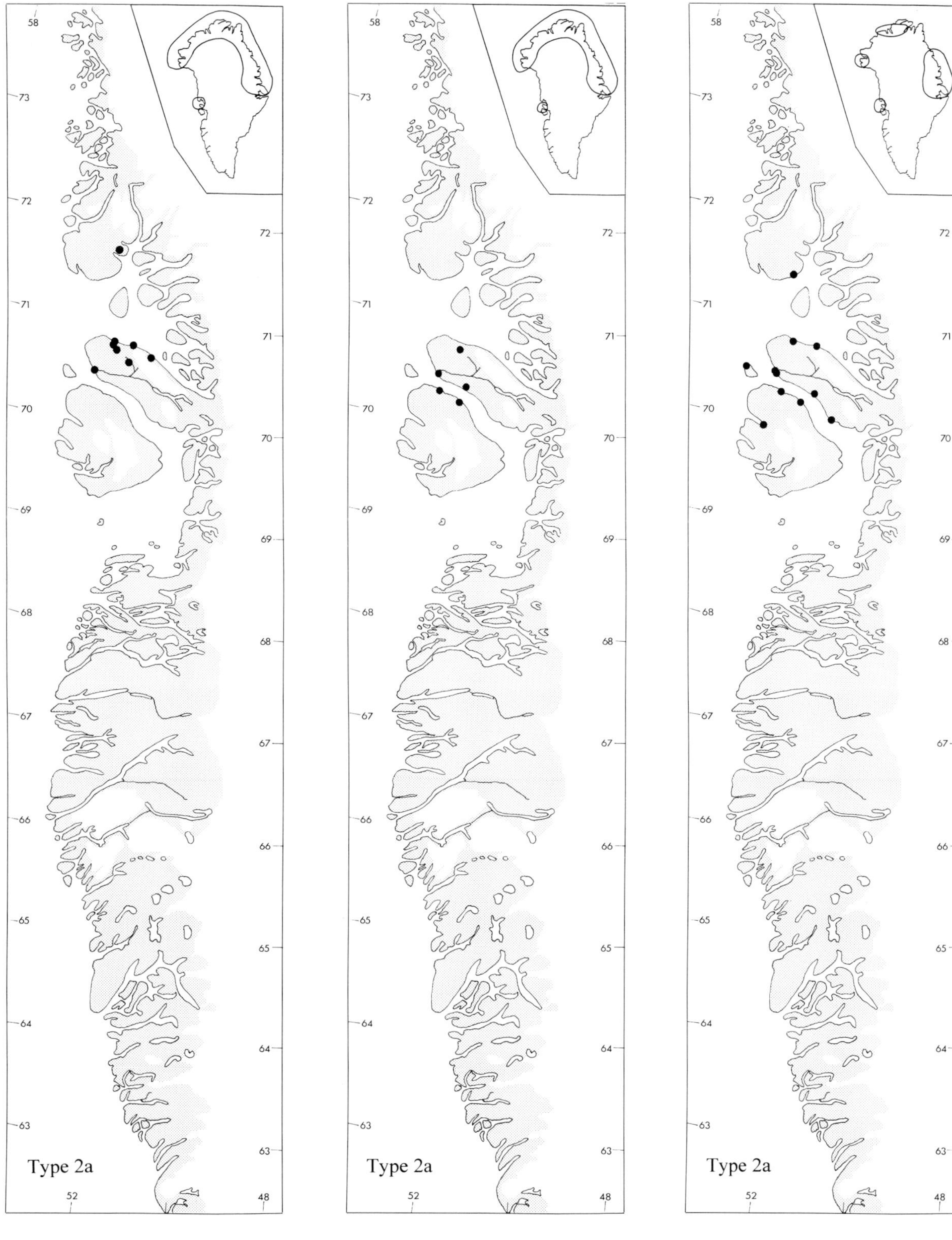

97. Poa pratensis var. colpodea

98. Potentilla rubricaulis

99. Puccinellia andersonii

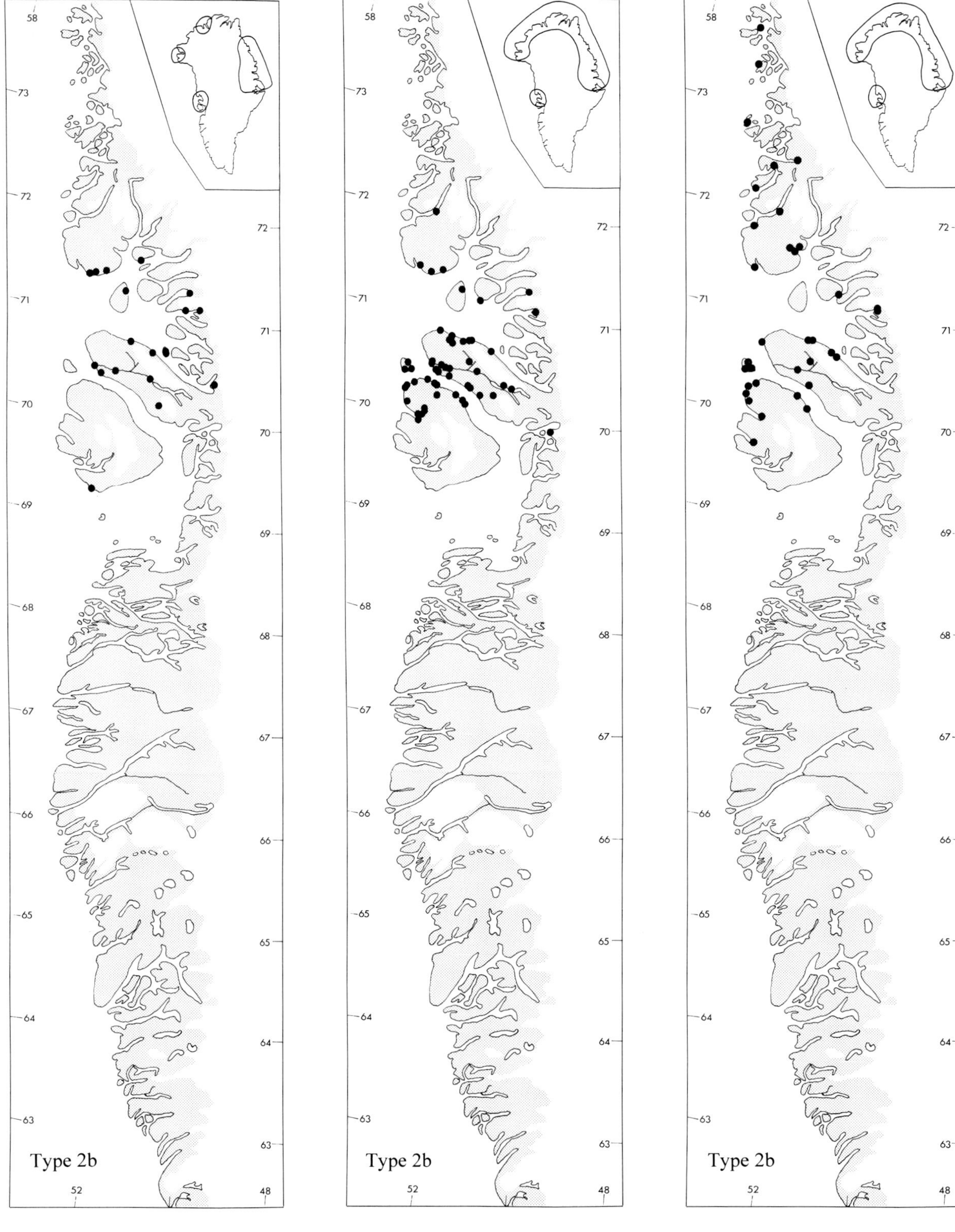

100. Carex atrofusca

101. Colpodium vahlianum

102. Eriophorum triste

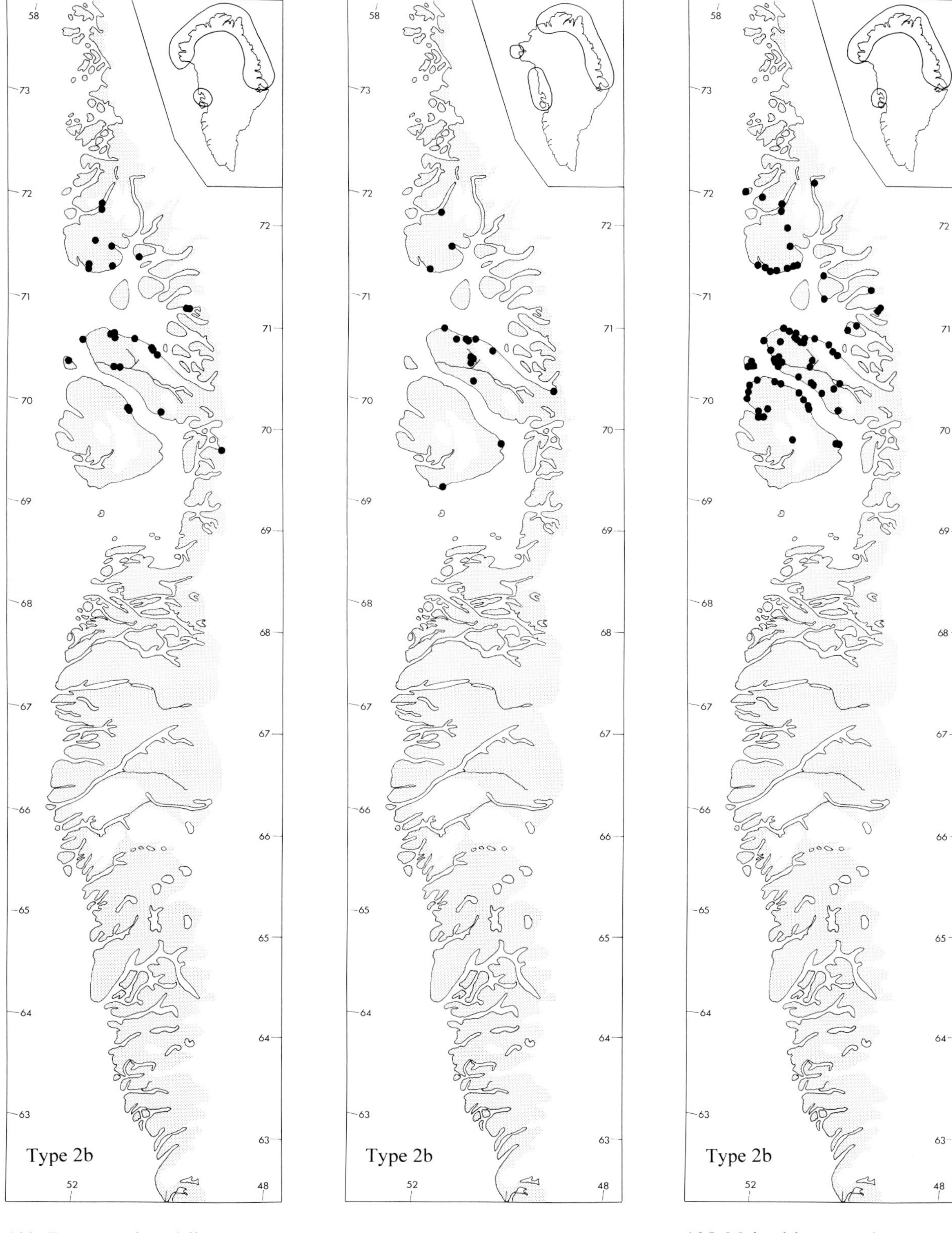

103. Eutrema edwardsii

104. Festuca hyperborea

105. Melandrium apetalum ssp. arcticum

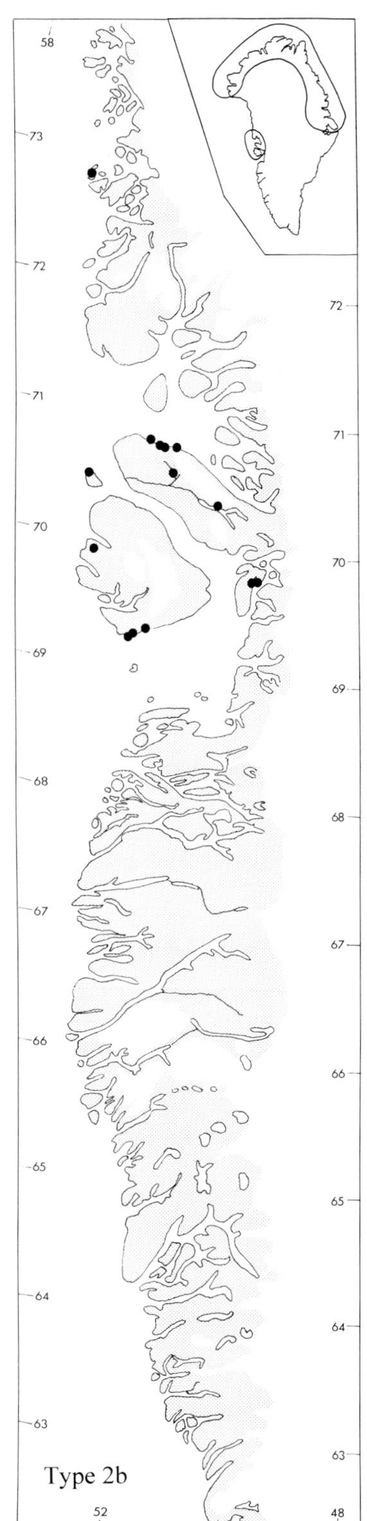

106. Phippsia algida
ssp. algidiformis

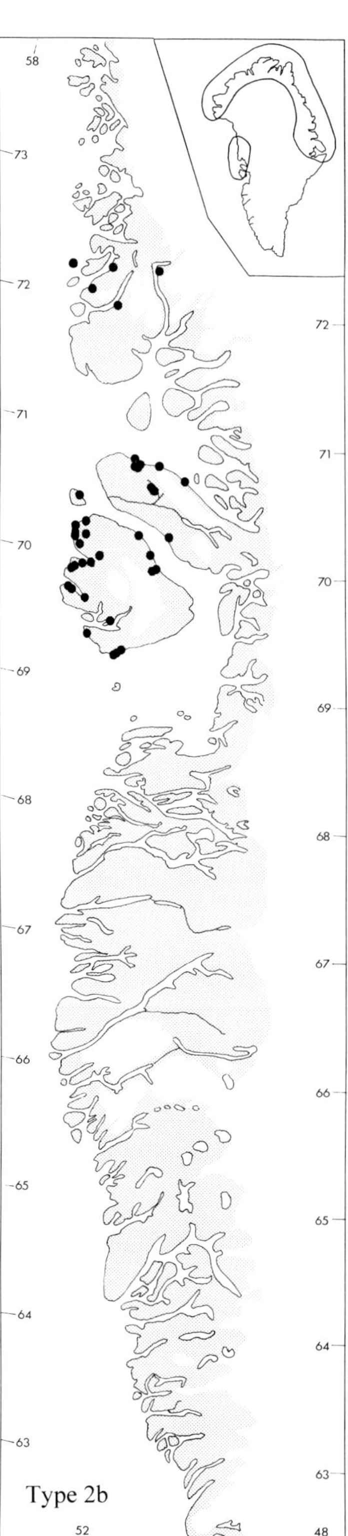

107. Ranunculus sulphureus

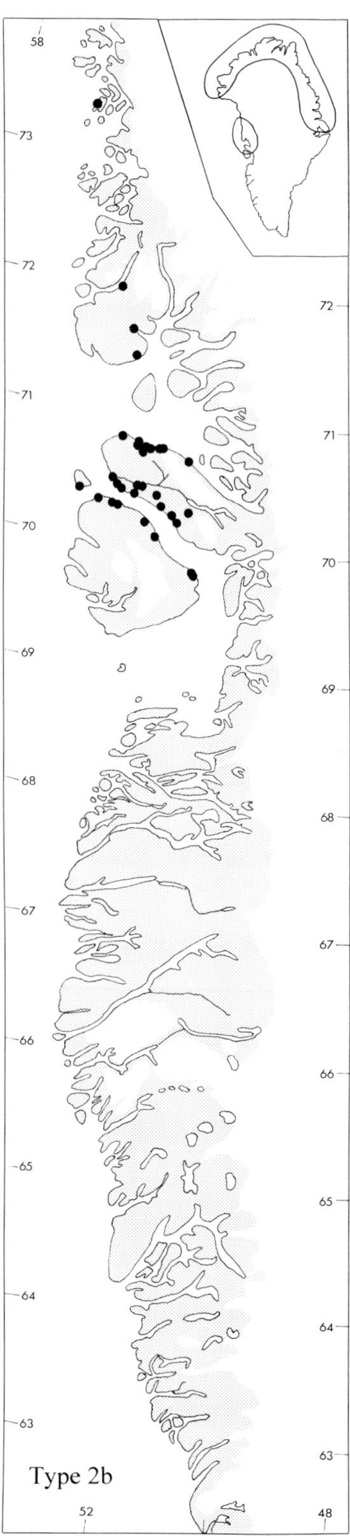

108. Taraxacum phymatocarpum

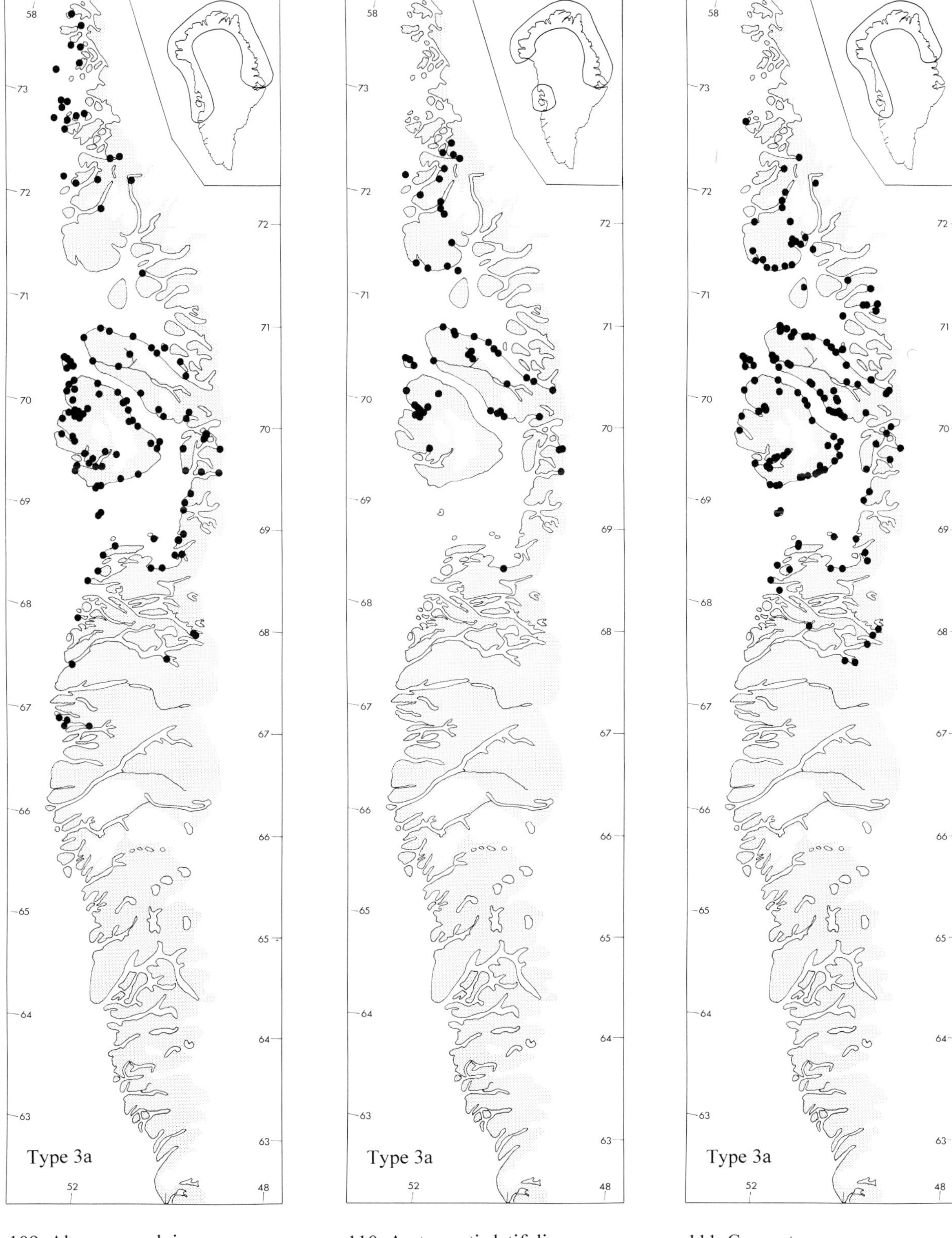

109. Alopecurus alpinus

110. Arctagrostis latifolia

111. Carex stans

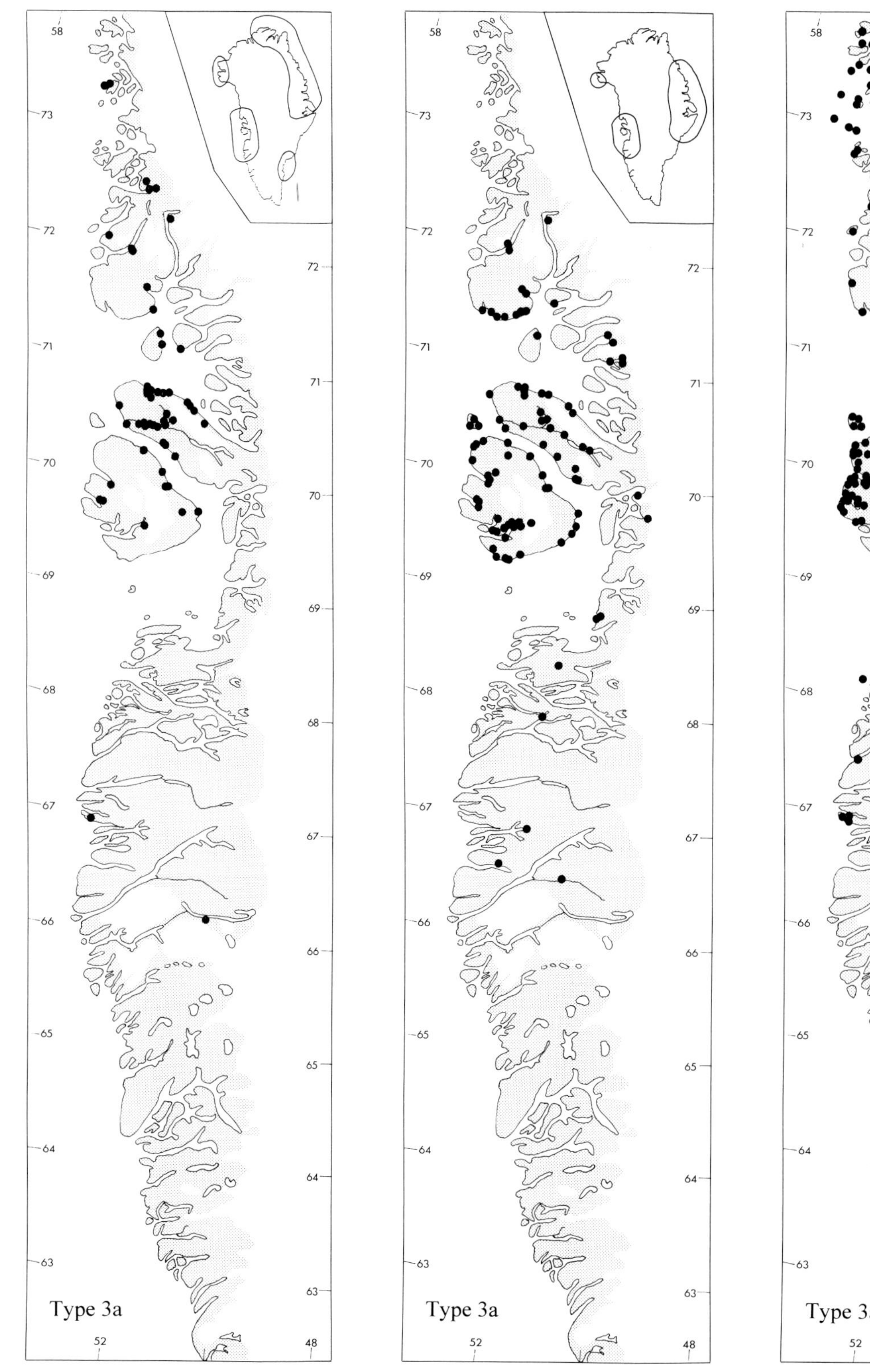

112. Erigeron eriocephalus

113. Minuartia stricta

114. Potentilla hyparctica

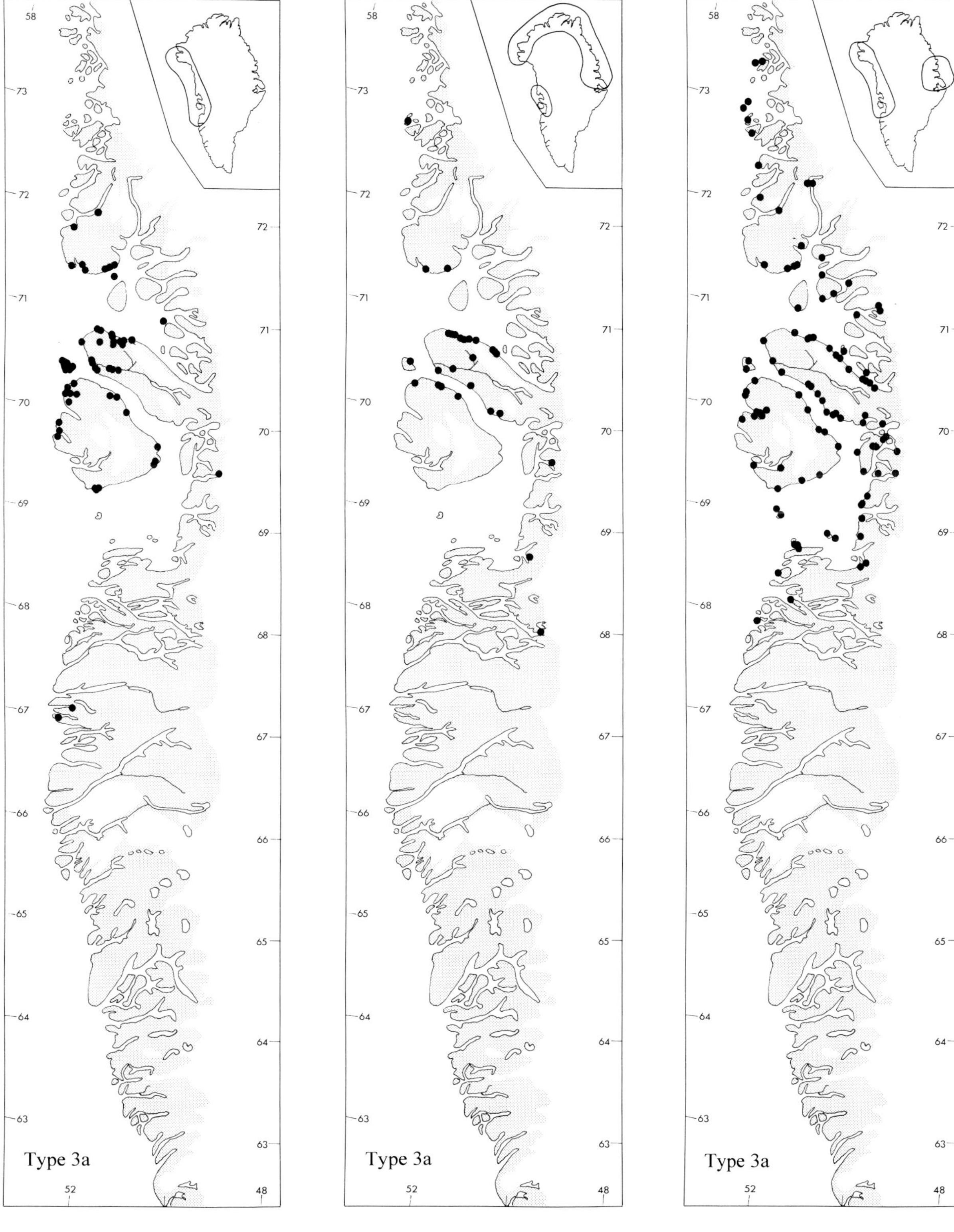

115. Potentilla vahliana

116. Puccinellia angustata

117. Puccinellia vaginata

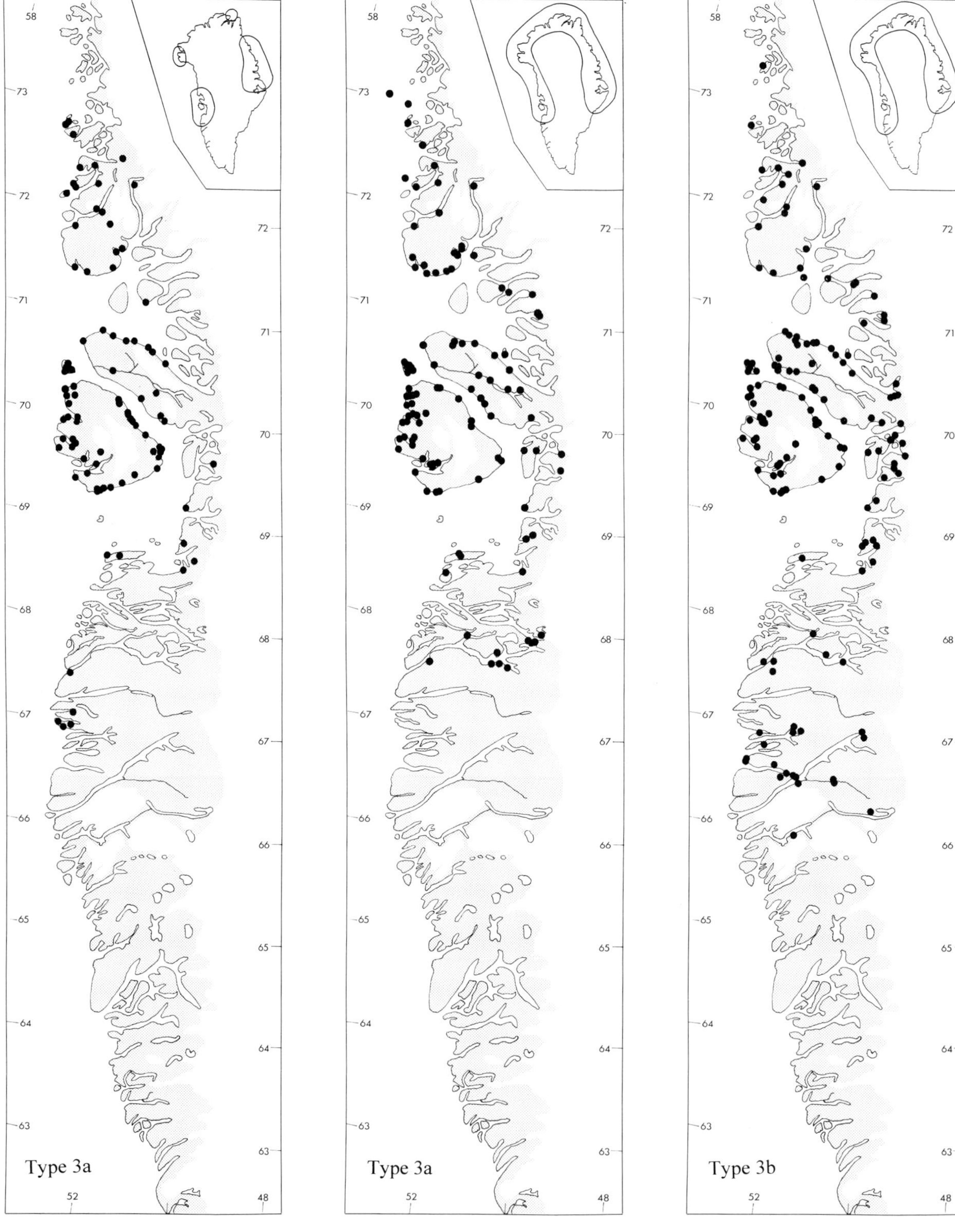

118. Ranunculus nivalis

119. Salix arctica

120. Carex misandra

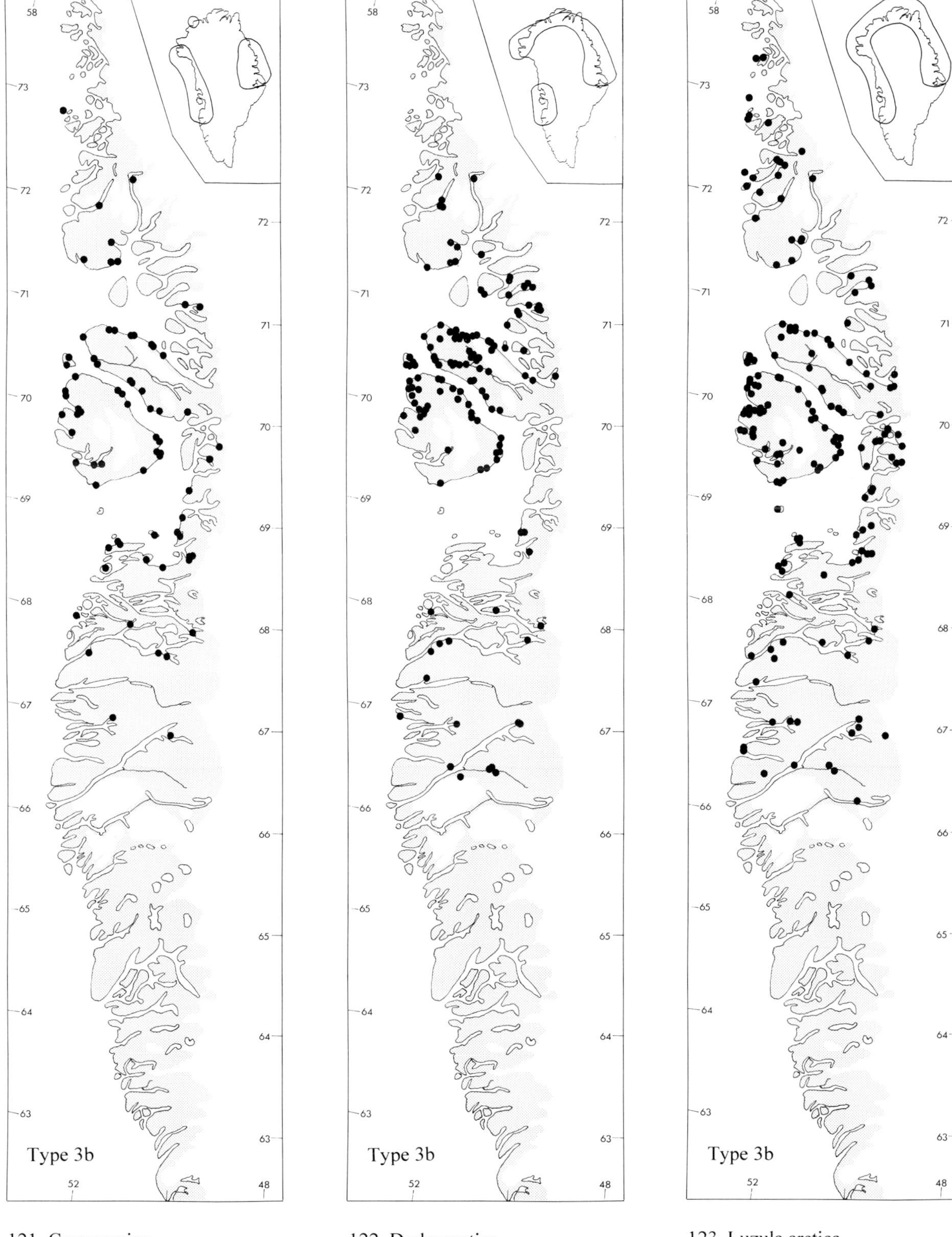

121. Carex ursina

122. Draba arctica (incl. D. groenlandica)

123. Luzula arctica

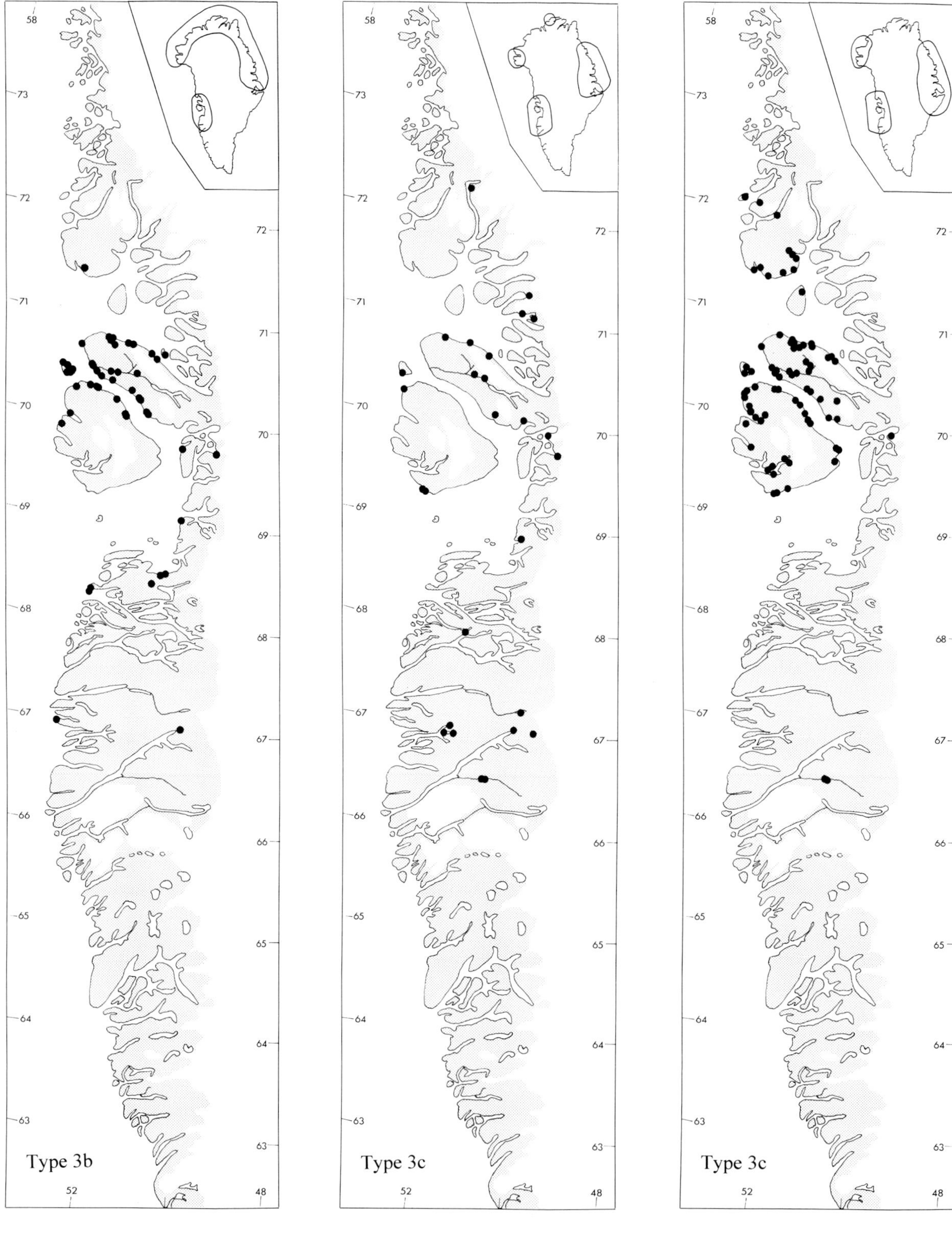

124. Potentilla pulchella

125. Carex marina

126. Draba alpina

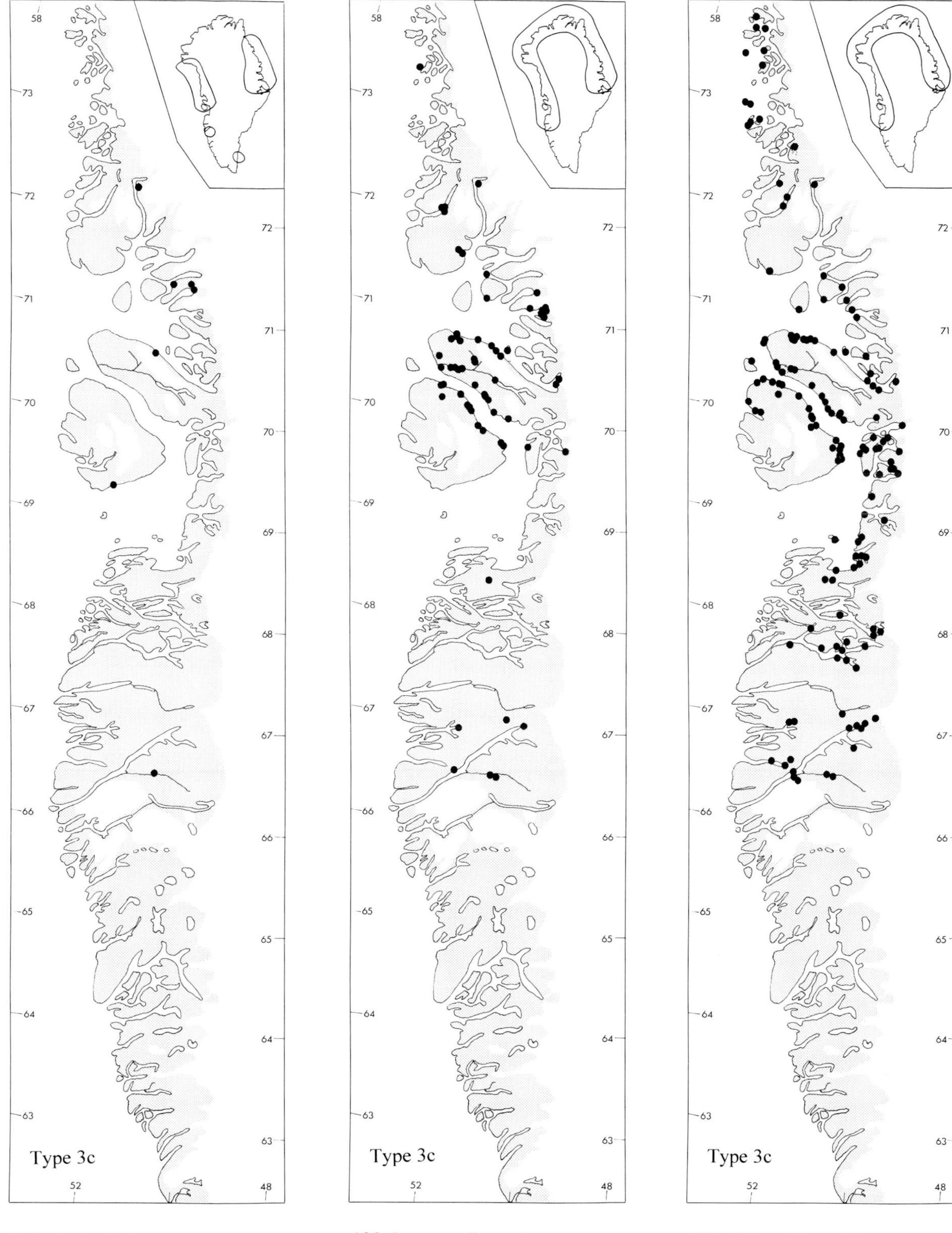

127. Draba fladnizensis

128. Lesquerella arctica

129. Melandrium triflorum

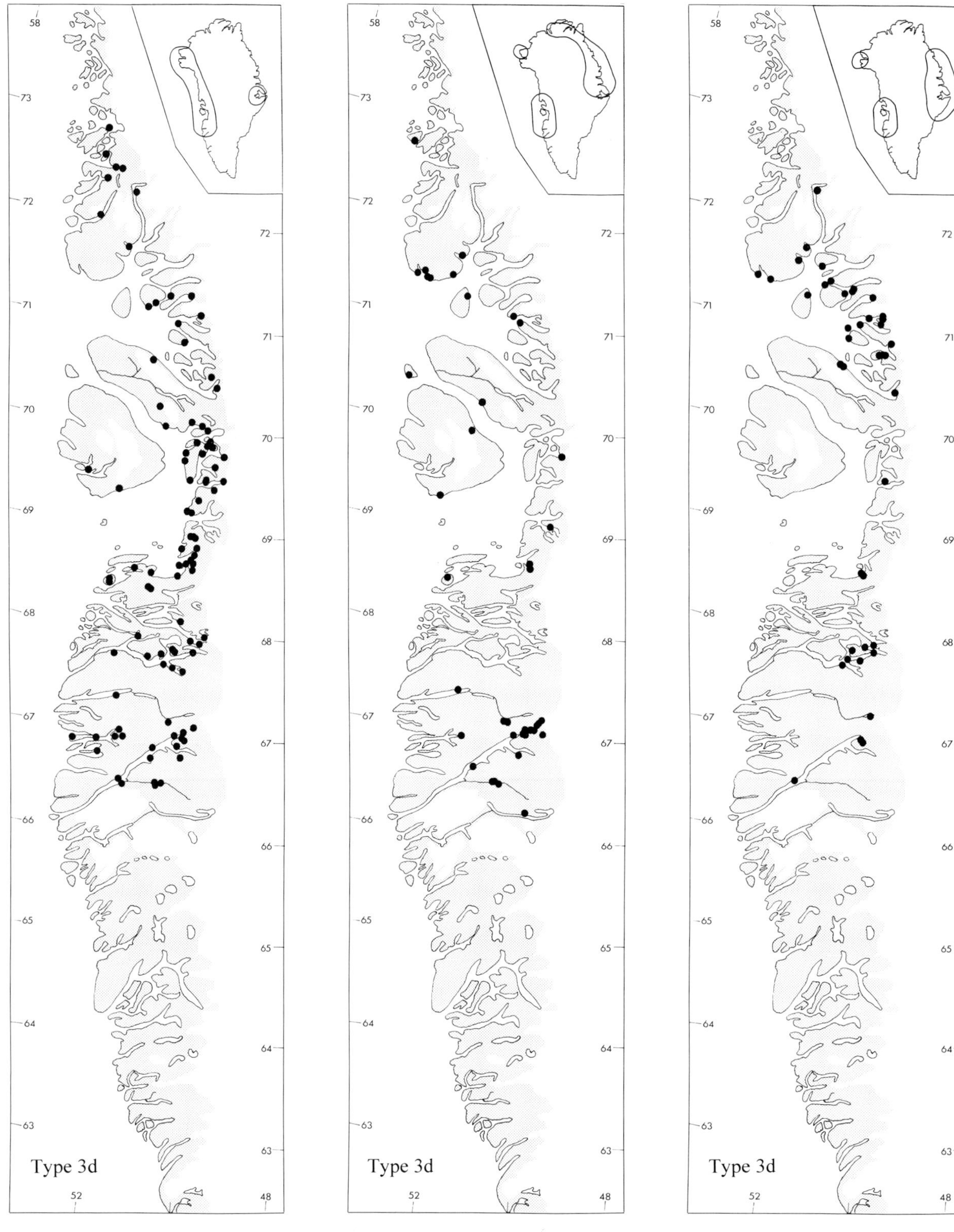

130. Dryopteris fragrans

131. Ranunculus affinis

132. Tofieldia coccinea

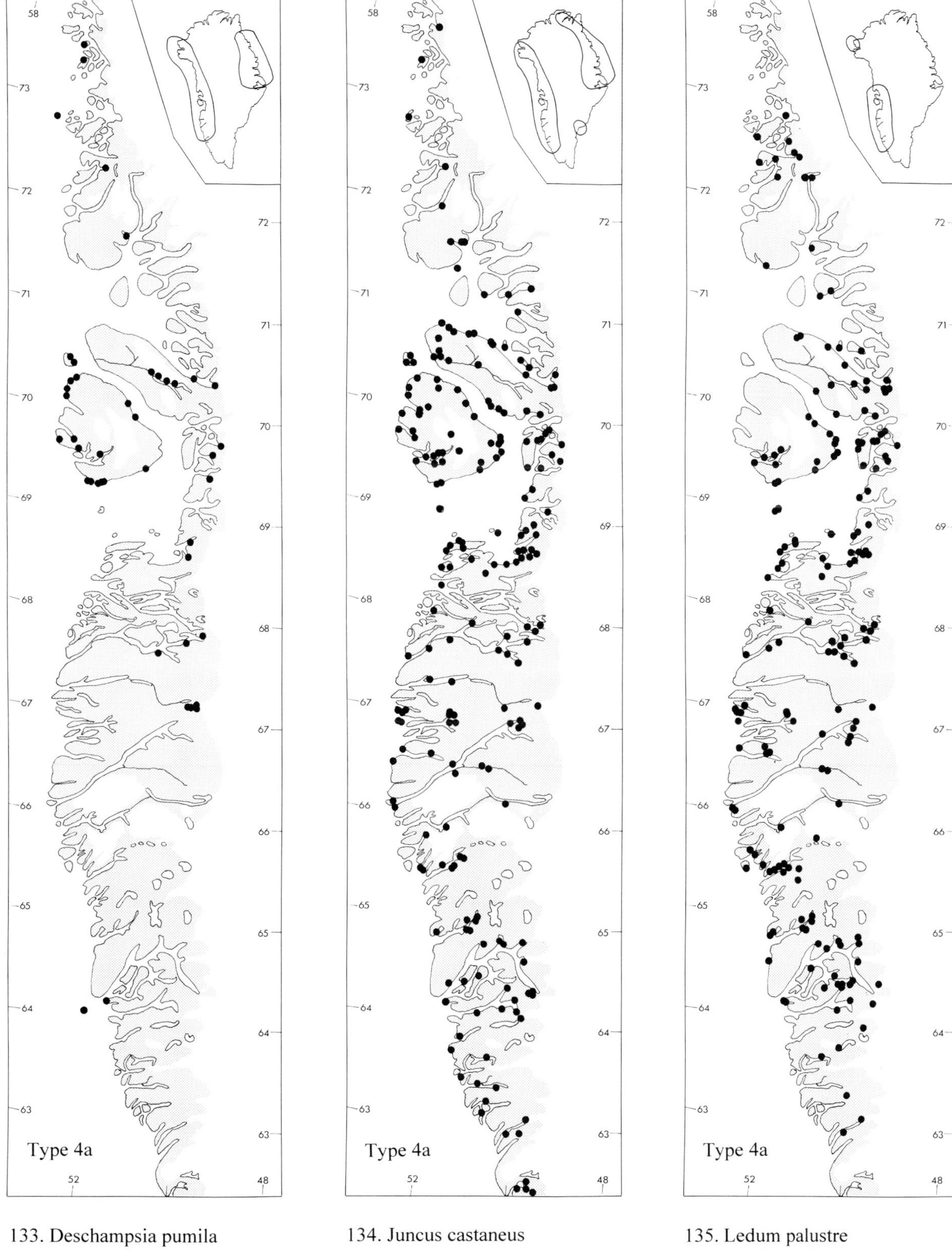

133. Deschampsia pumila

134. Juncus castaneus

135. Ledum palustre ssp. decumbens

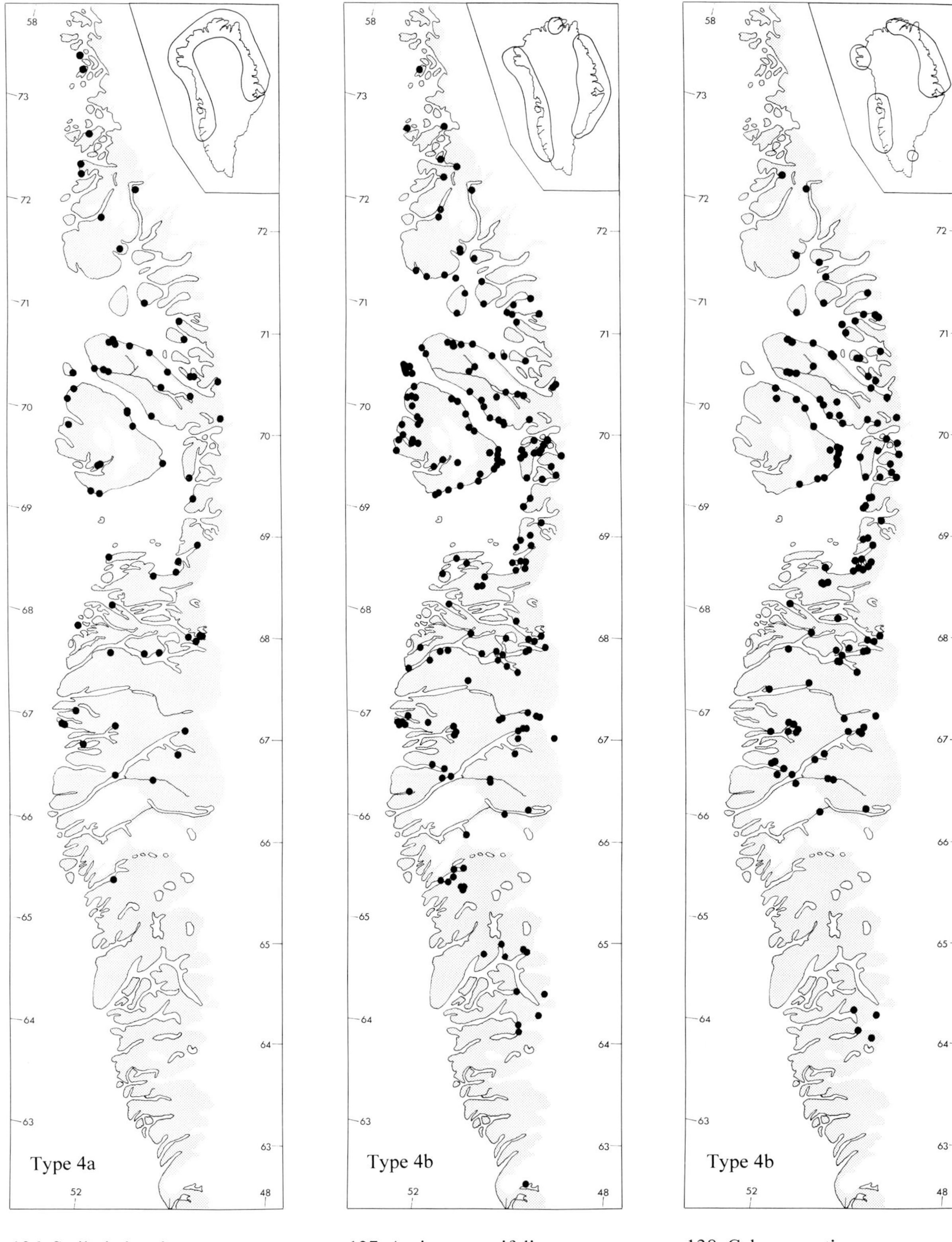

136. Stellaria longipes
(B: sepals ciliate)

137. Arnica angustifolia

138. Calamagrostis purpurascens

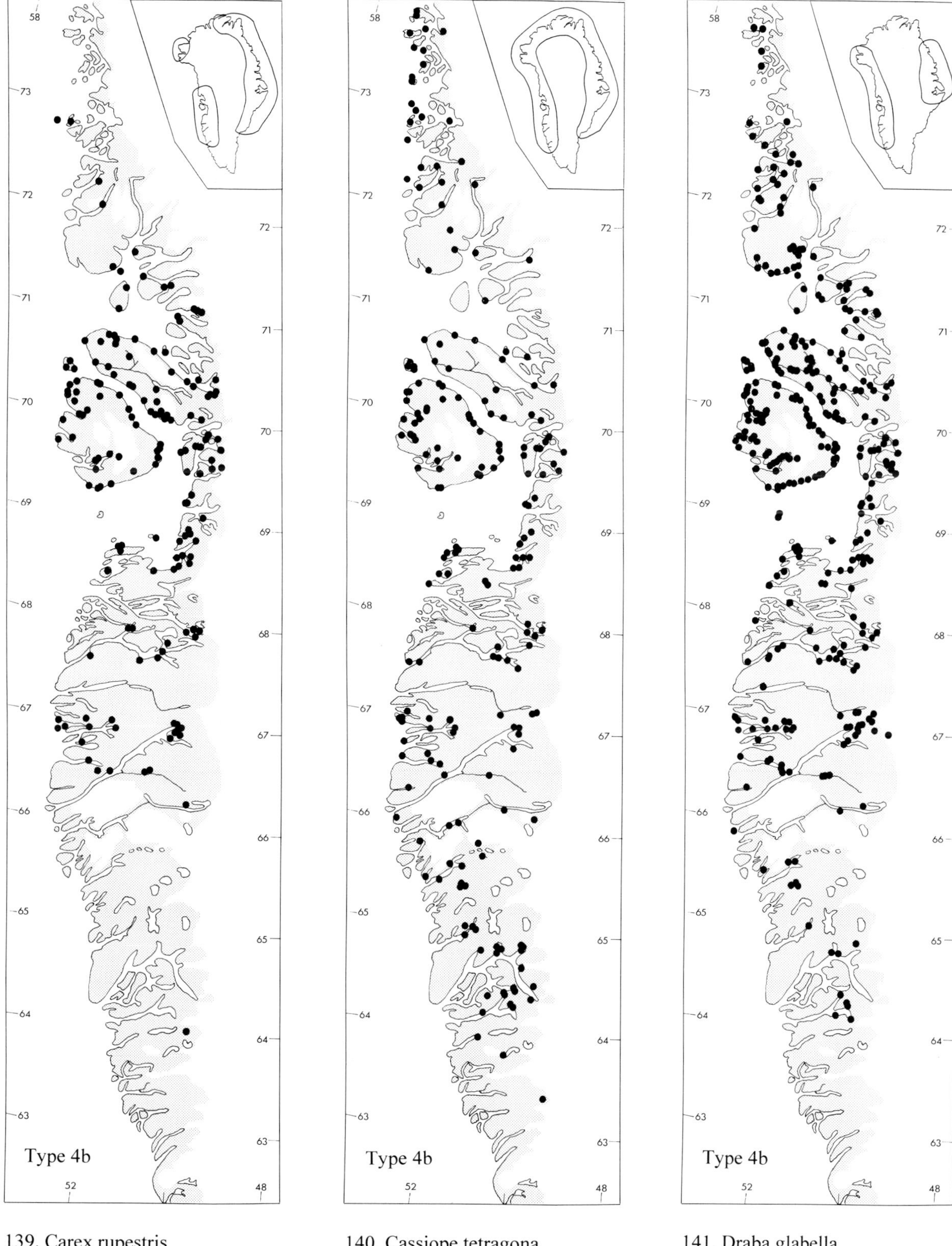

139. Carex rupestris

140. Cassiope tetragona

141. Draba glabella

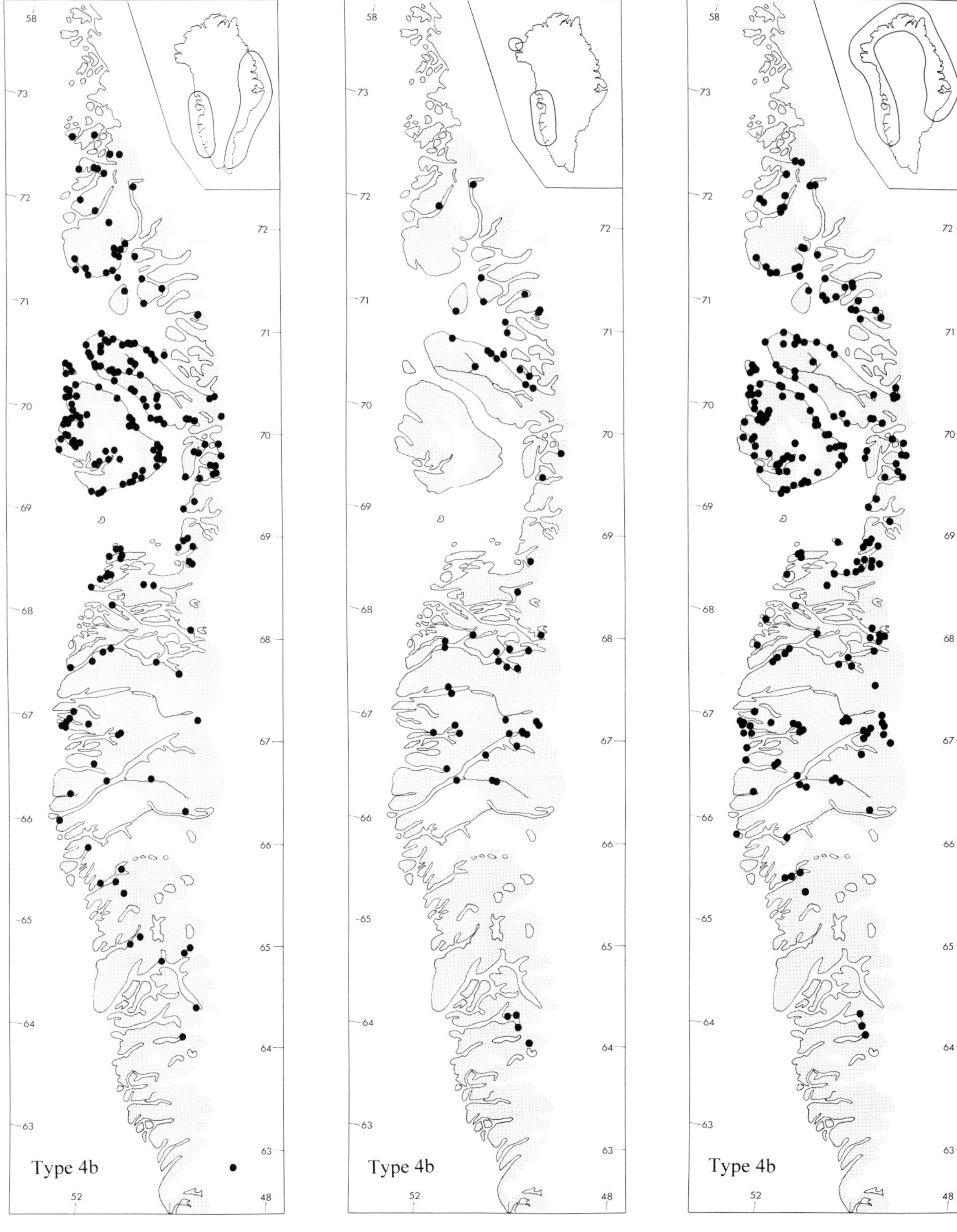

142. Erigeron humilis

143. Halimolobus mollis

144. Melandrium affine

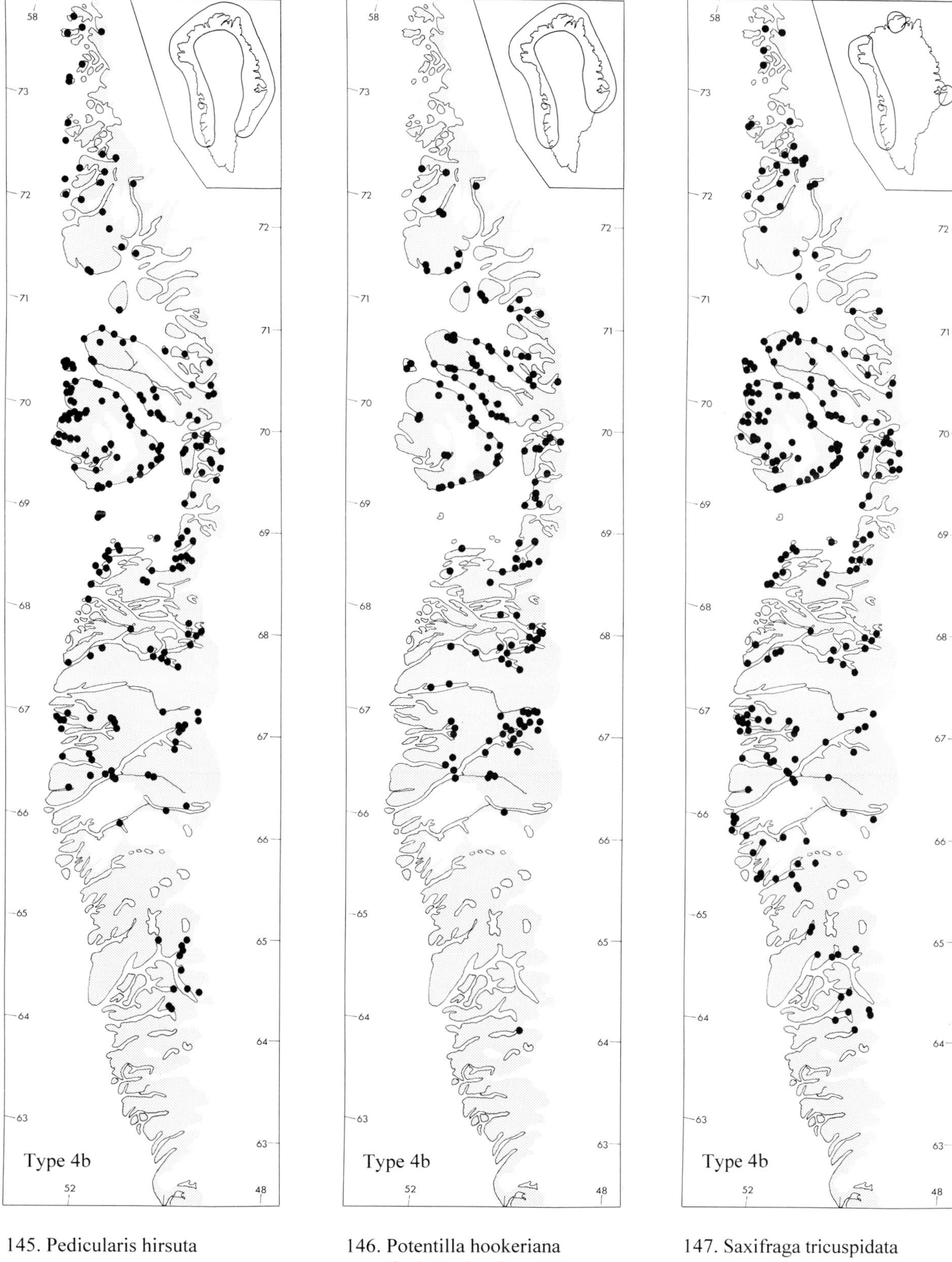

145. Pedicularis hirsuta

146. Potentilla hookeriana (incl. P. chamissonis)

147. Saxifraga tricuspidata

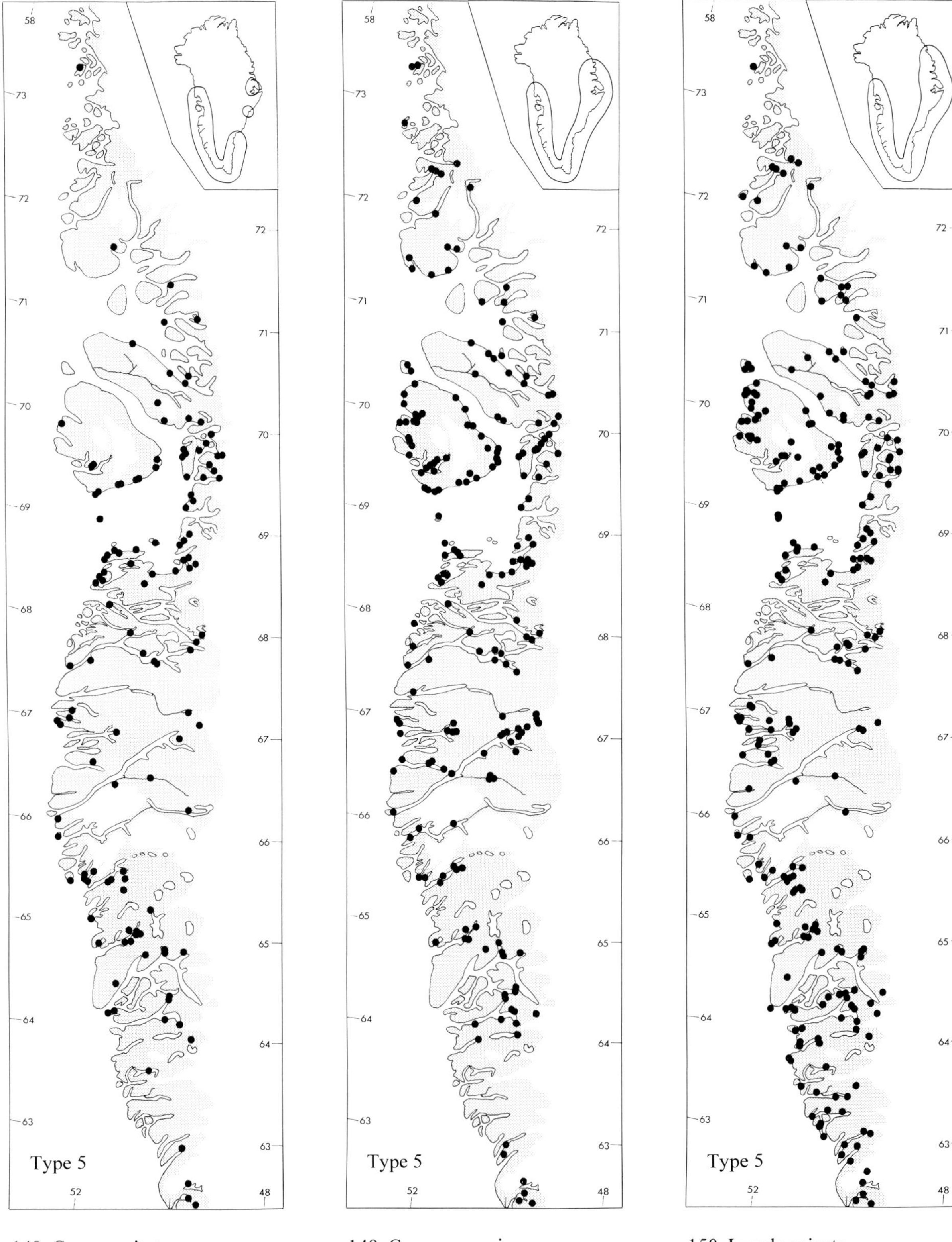

148. Carex capitata
ssp. arctogena

149. Carex norvegica

150. Luzula spicata

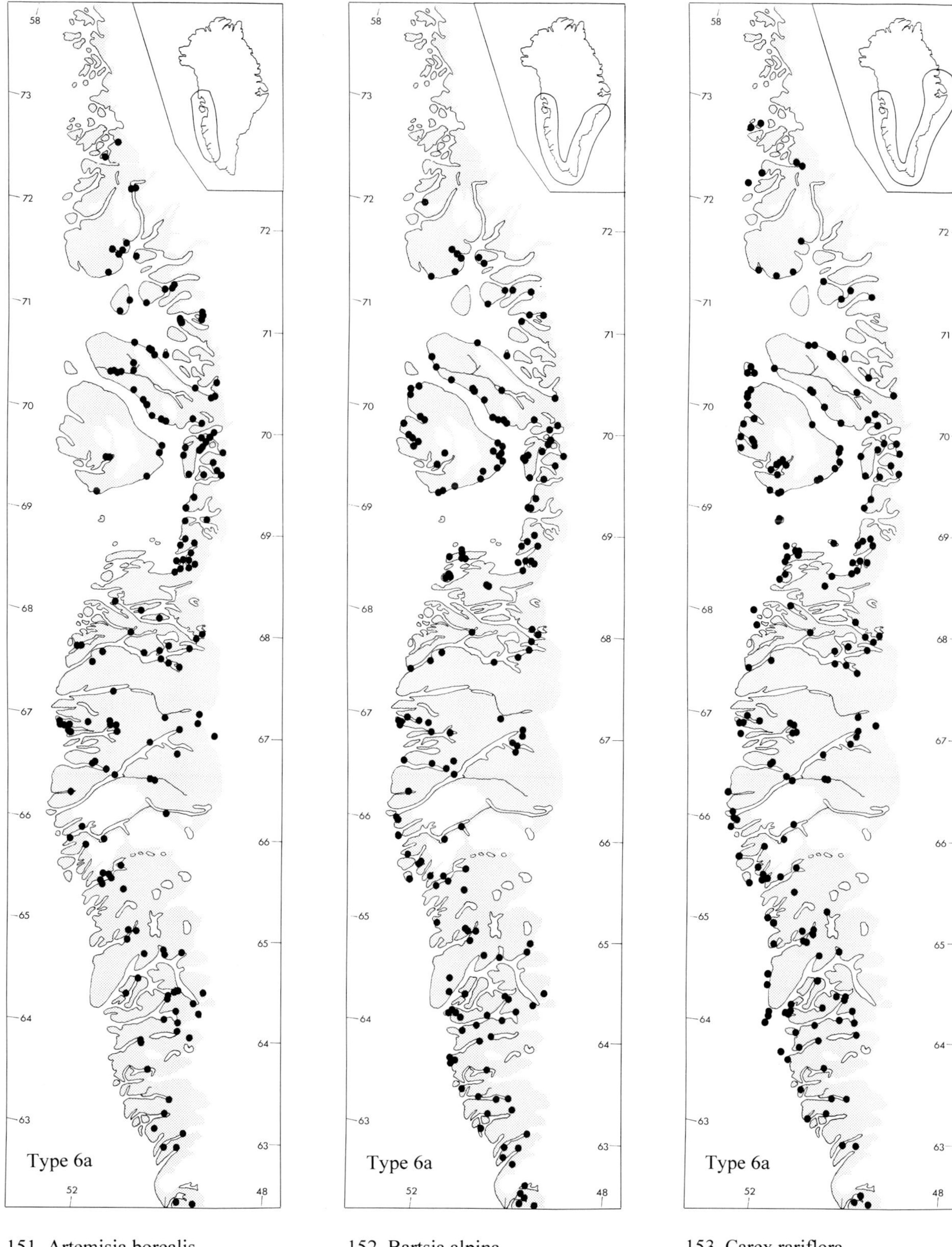

151. Artemisia borealis

152. Bartsia alpina

153. Carex rariflora

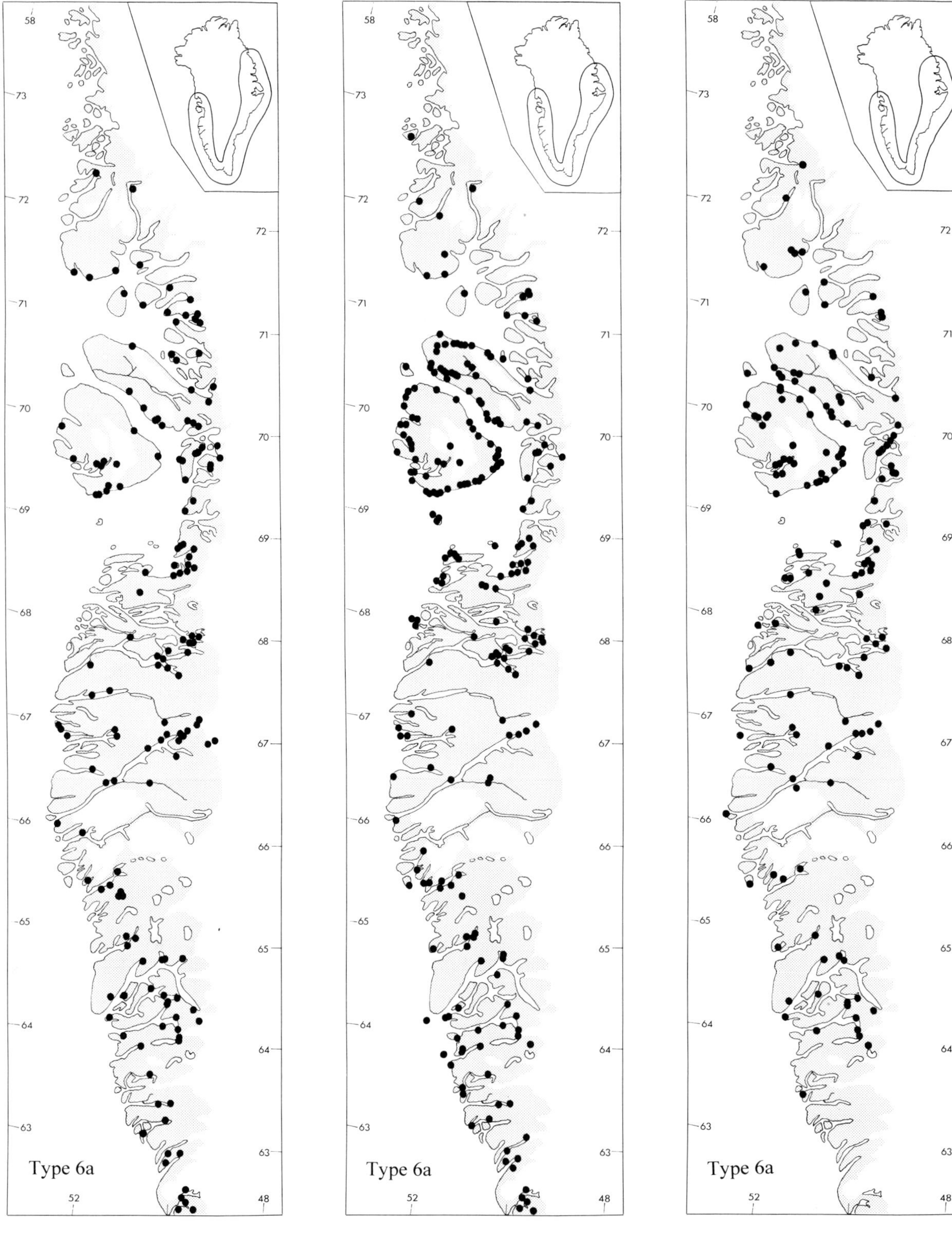

154. Euphrasia frigida

155. Festuca rubra

156. Juncus arcticus

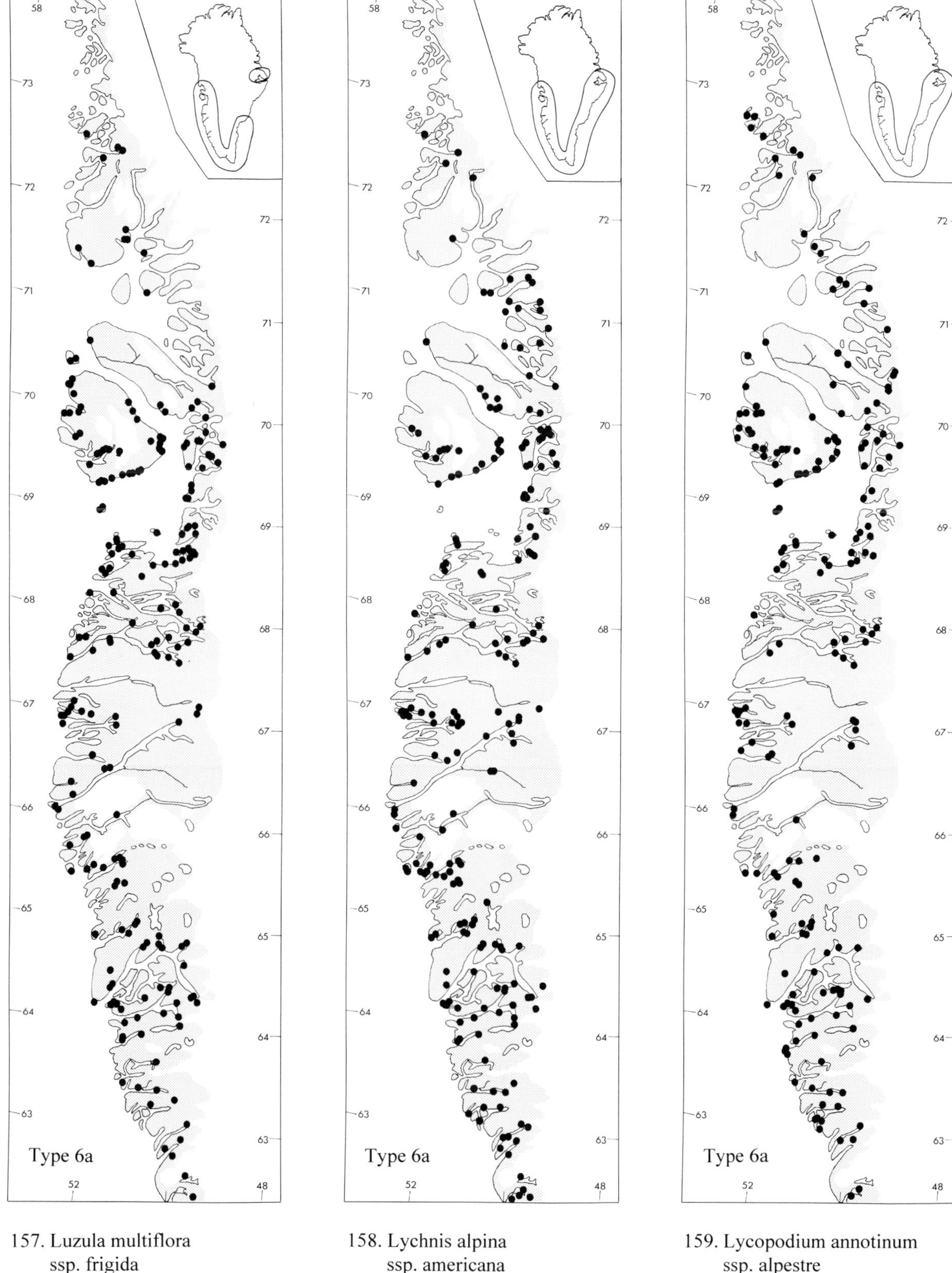

157. Luzula multiflora ssp. frigida

158. Lychnis alpina ssp. americana

159. Lycopodium annotinum ssp. alpestre

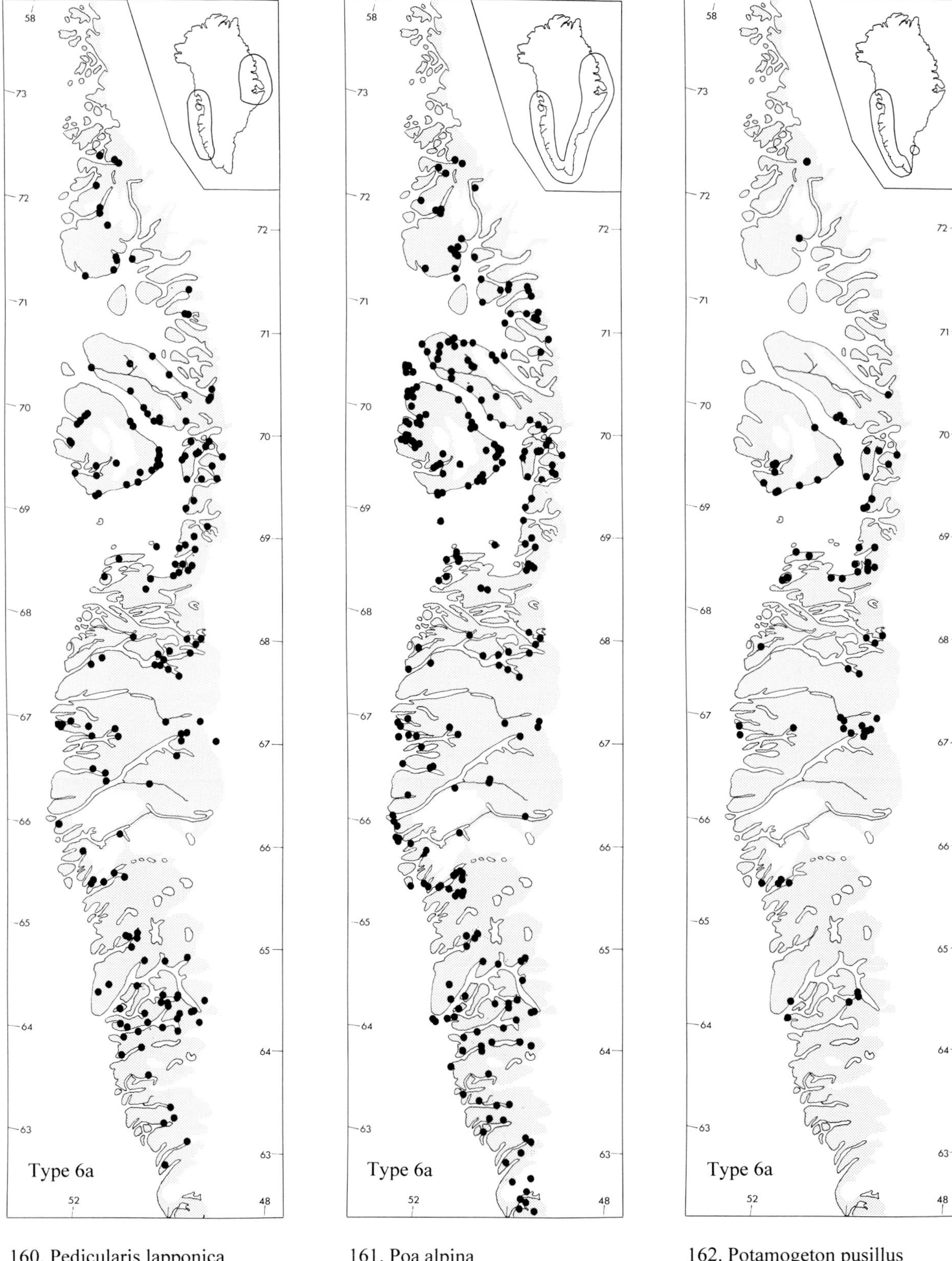

160. Pedicularis lapponica

161. Poa alpina

162. Potamogeton pusillus ssp. groenlandicus

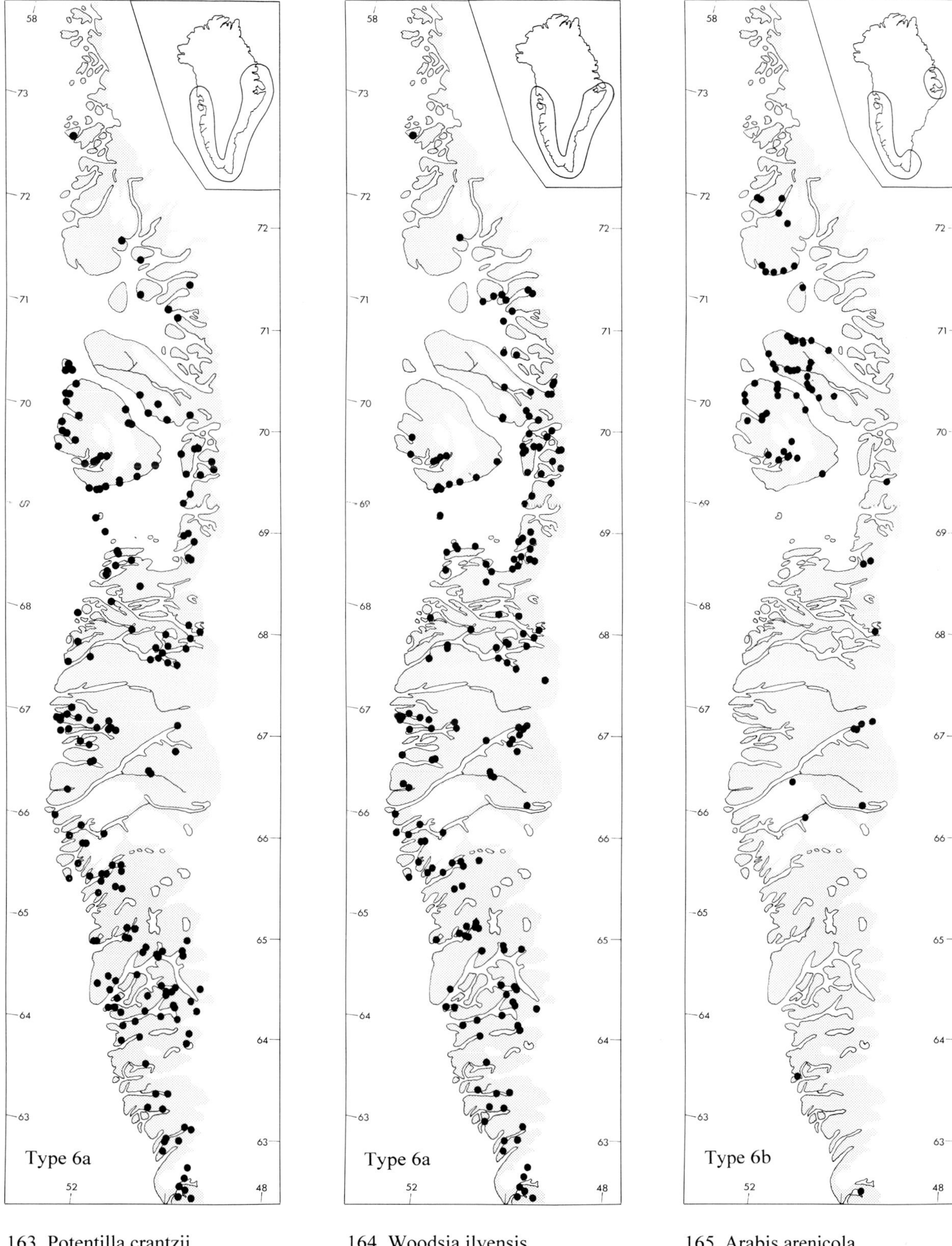

163. Potentilla crantzii

164. Woodsia ilvensis

165. Arabis arenicola

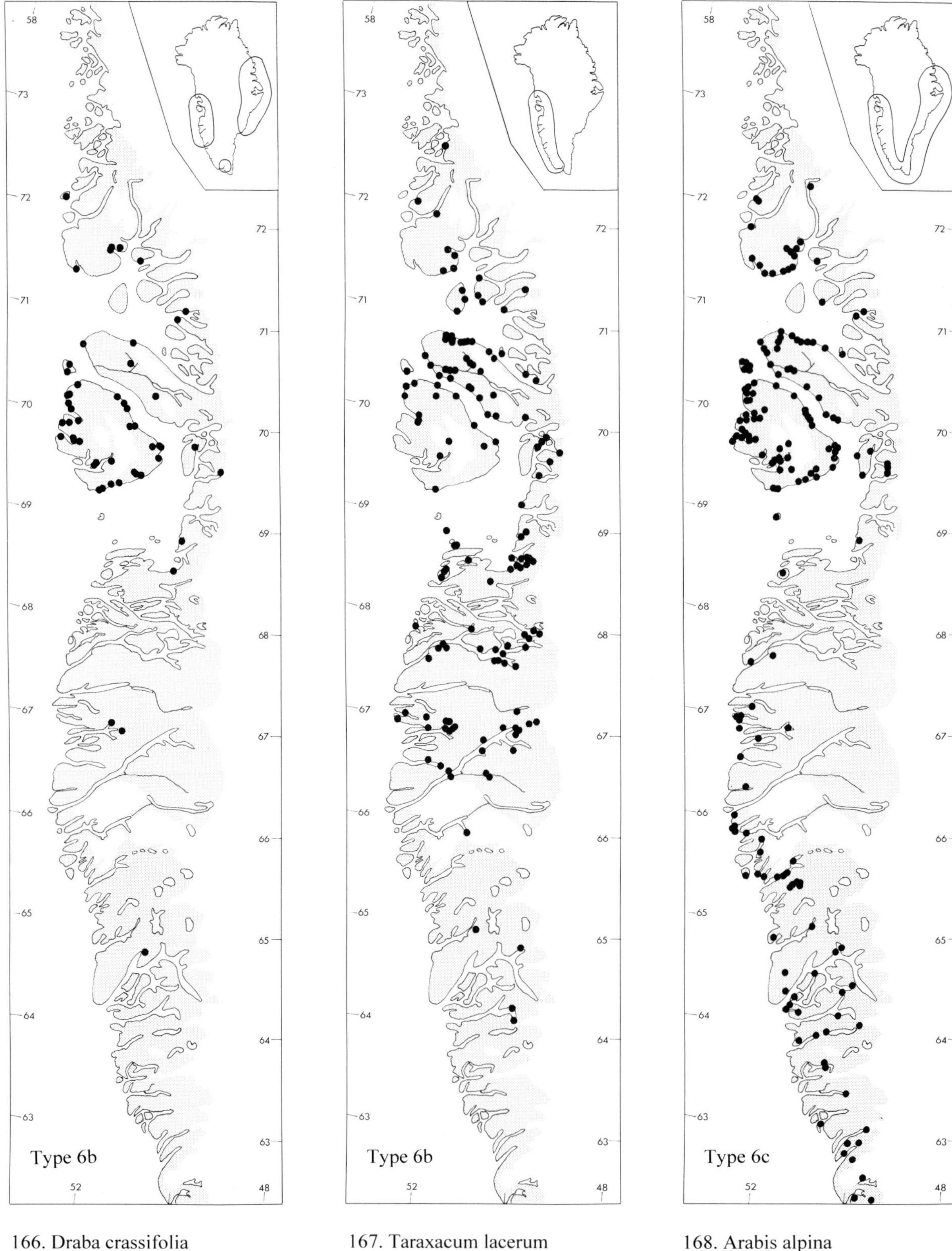

166. Draba crassifolia

167. Taraxacum lacerum (incl. T. umbrinum)

168. Arabis alpina

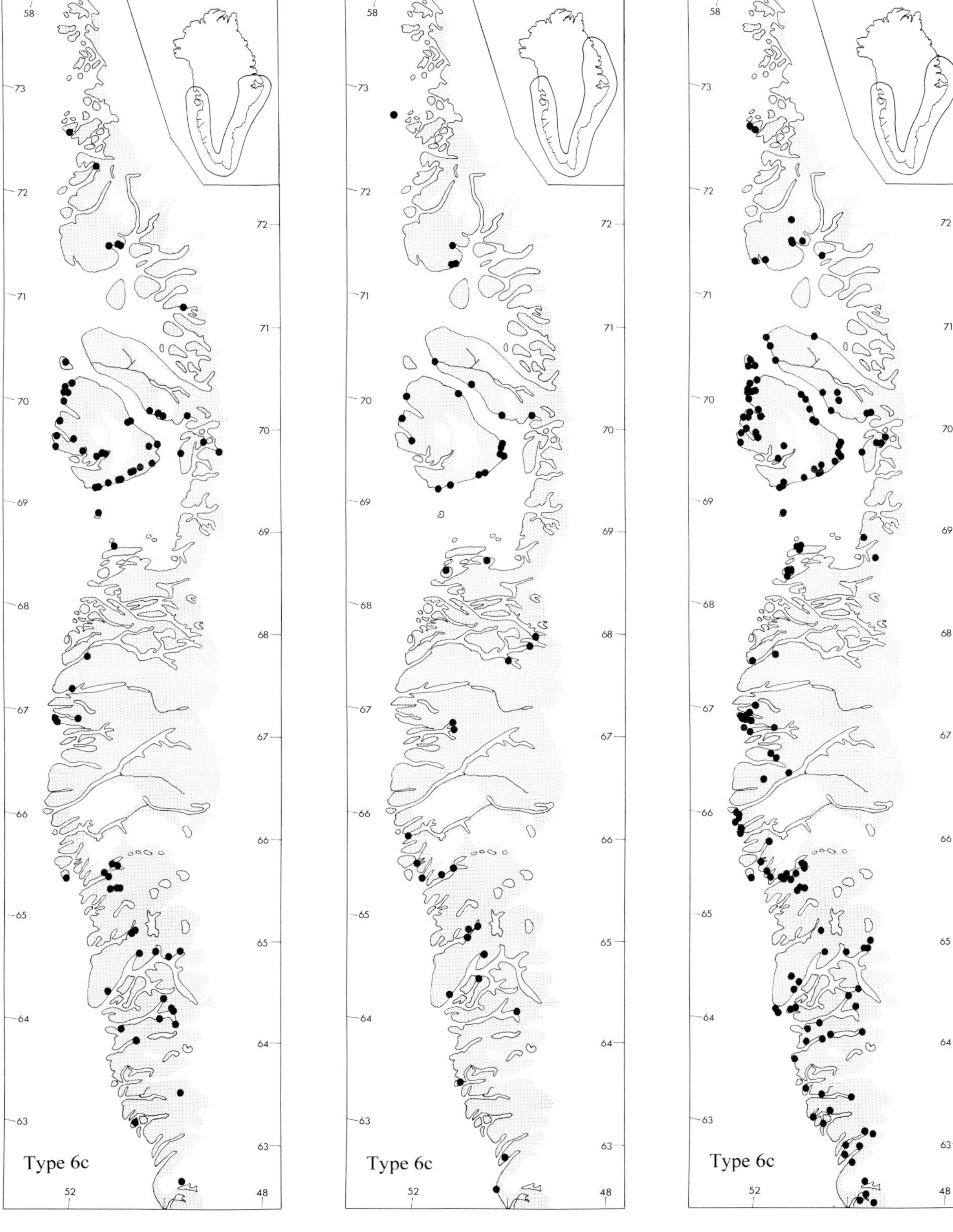

169. Carex macloviana

170. Carex subspathacea

171. Cerastium cerastoides

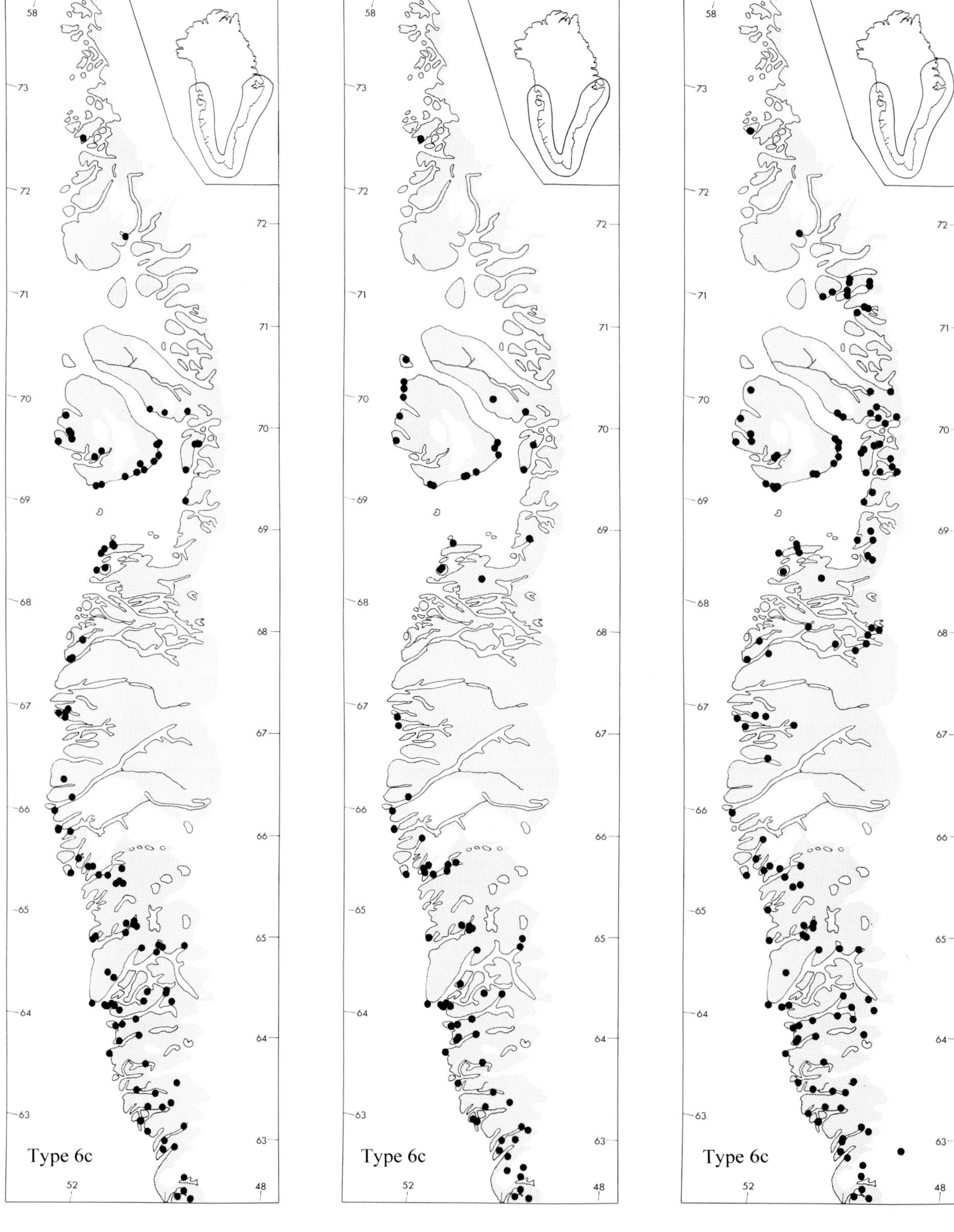

172. Diphasiastrum alpinum

173. Gnaphalium supinum

174. Juncus trifidus

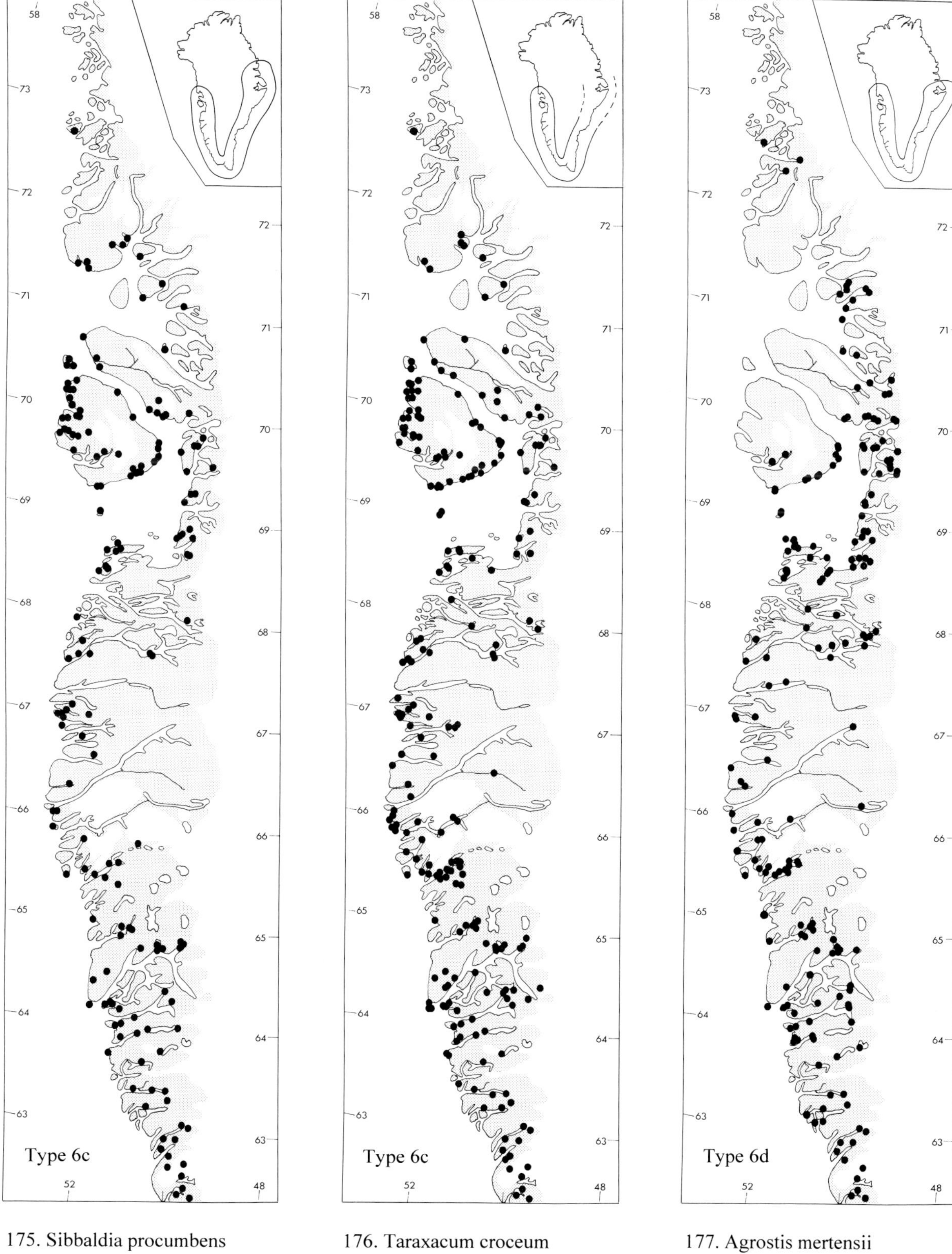

175. Sibbaldia procumbens

176. Taraxacum croceum (incl. T. amphiphron)

177. Agrostis mertensii

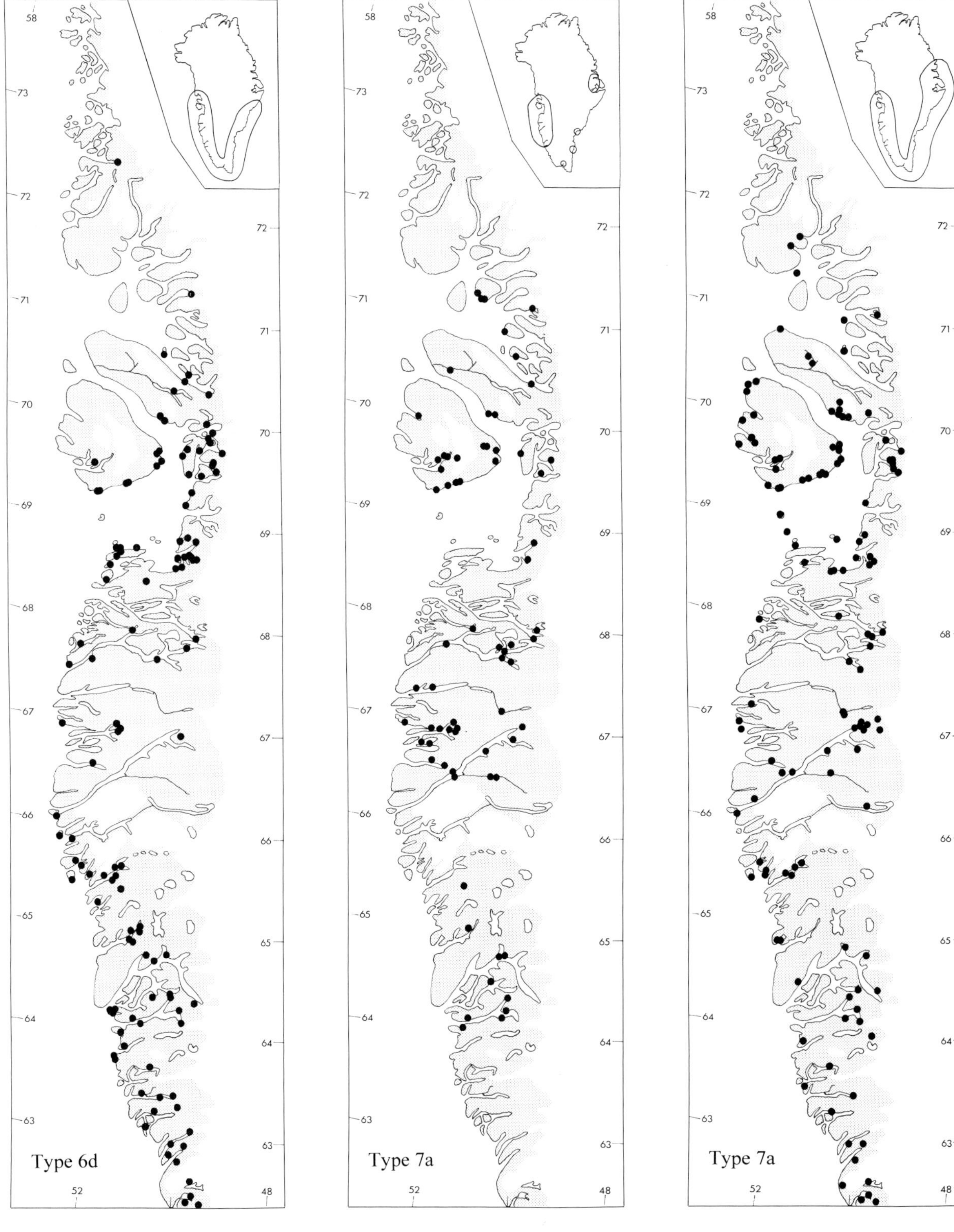

178. Scirpus caespitosus

179. Arabis holboellii

180. Calamagrostis neglecta

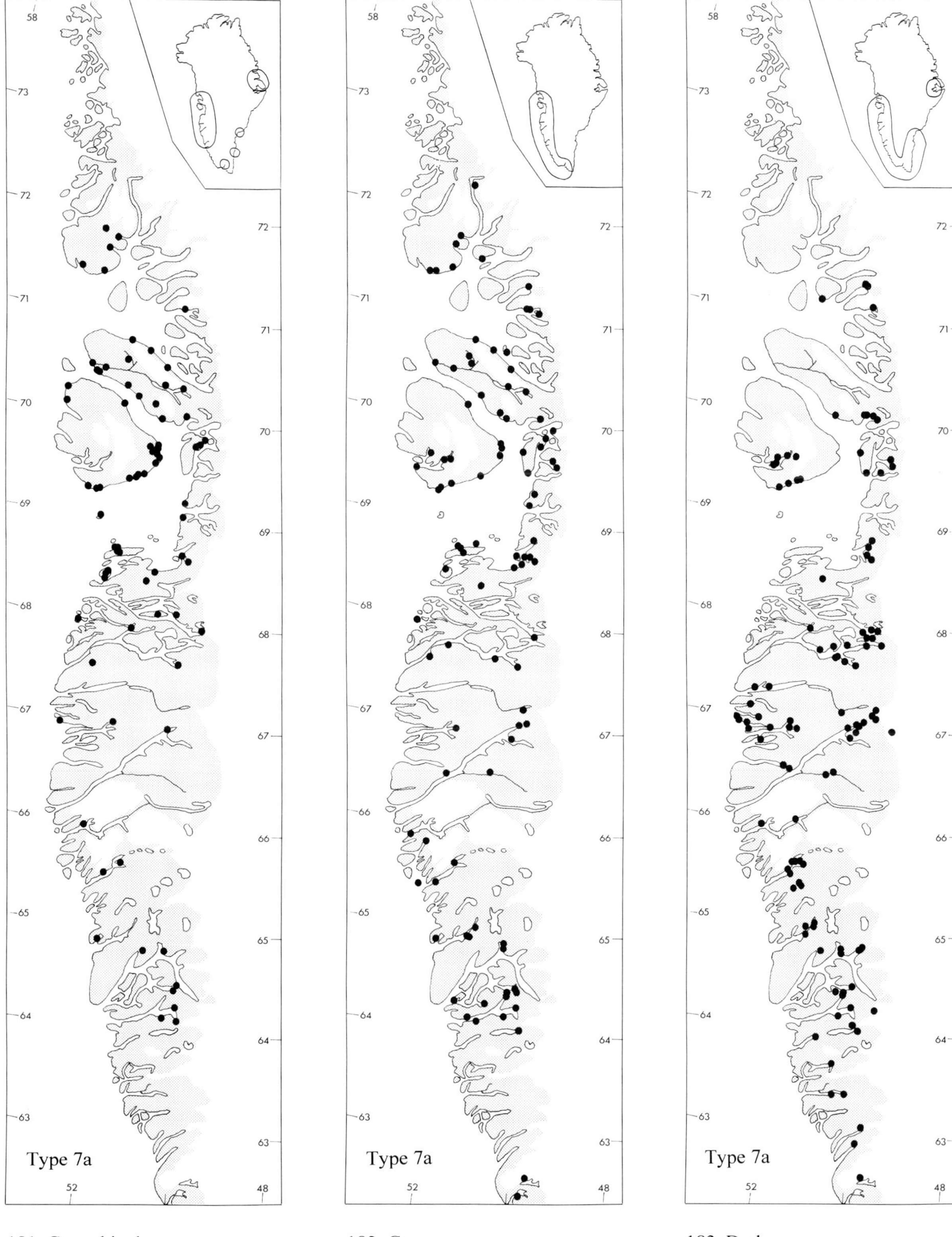

181. Carex bicolor

182. Carex gynocrates

183. Draba aurea

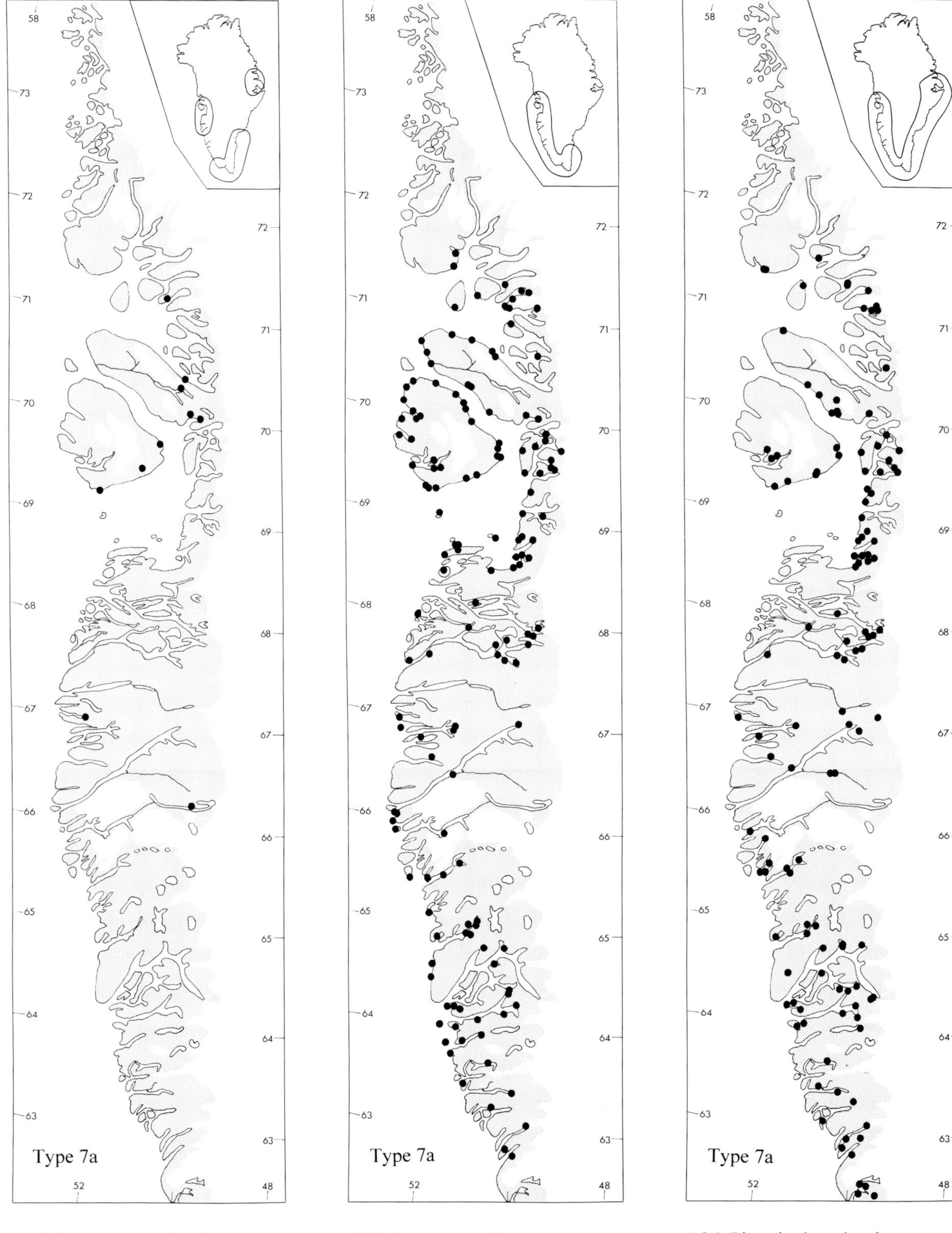

184. Festuca groenlandica

185. Leymus mollis

186. Pinguicula vulgaris

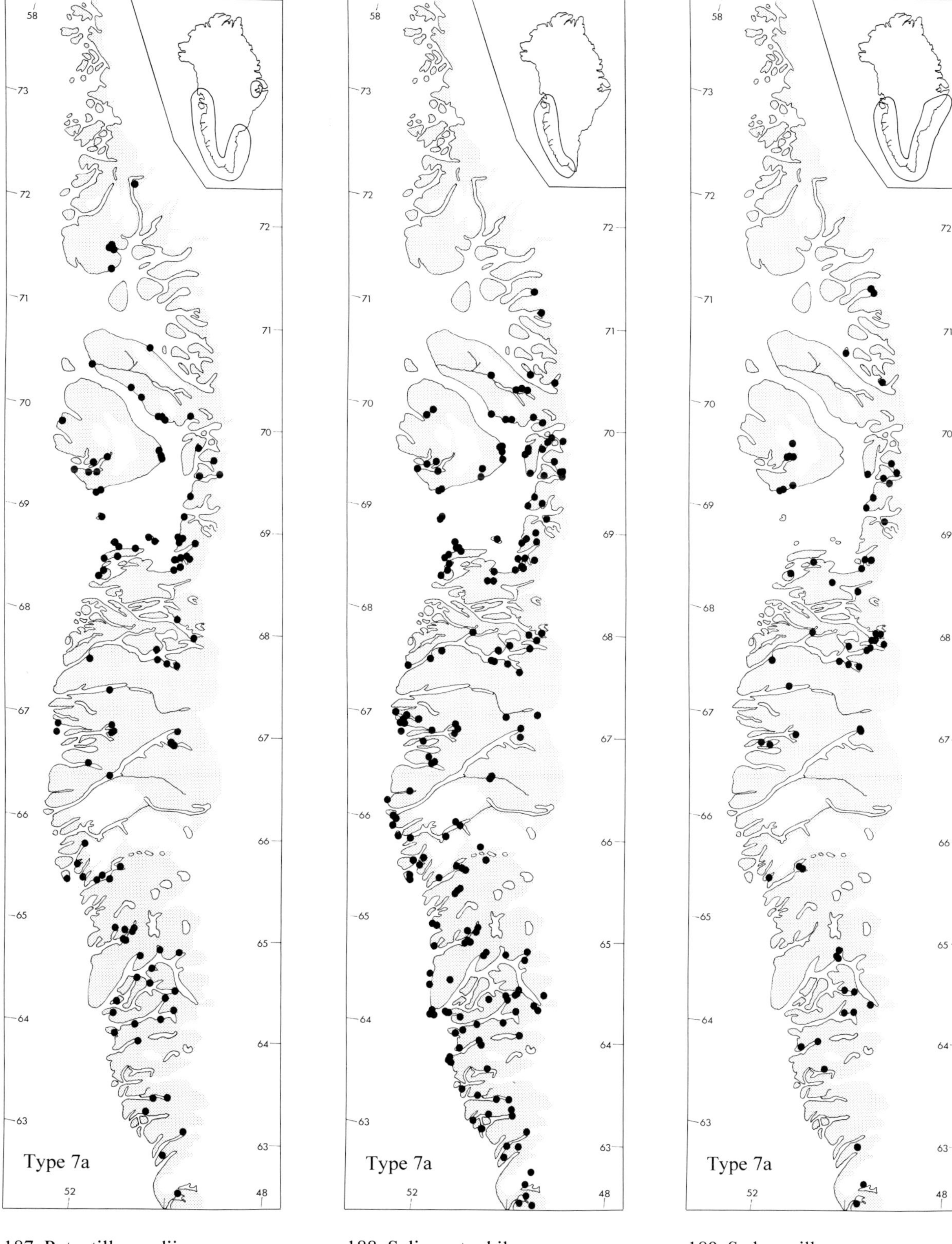

187. Potentilla egedii

188. Salix arctophila

189. Sedum villosum

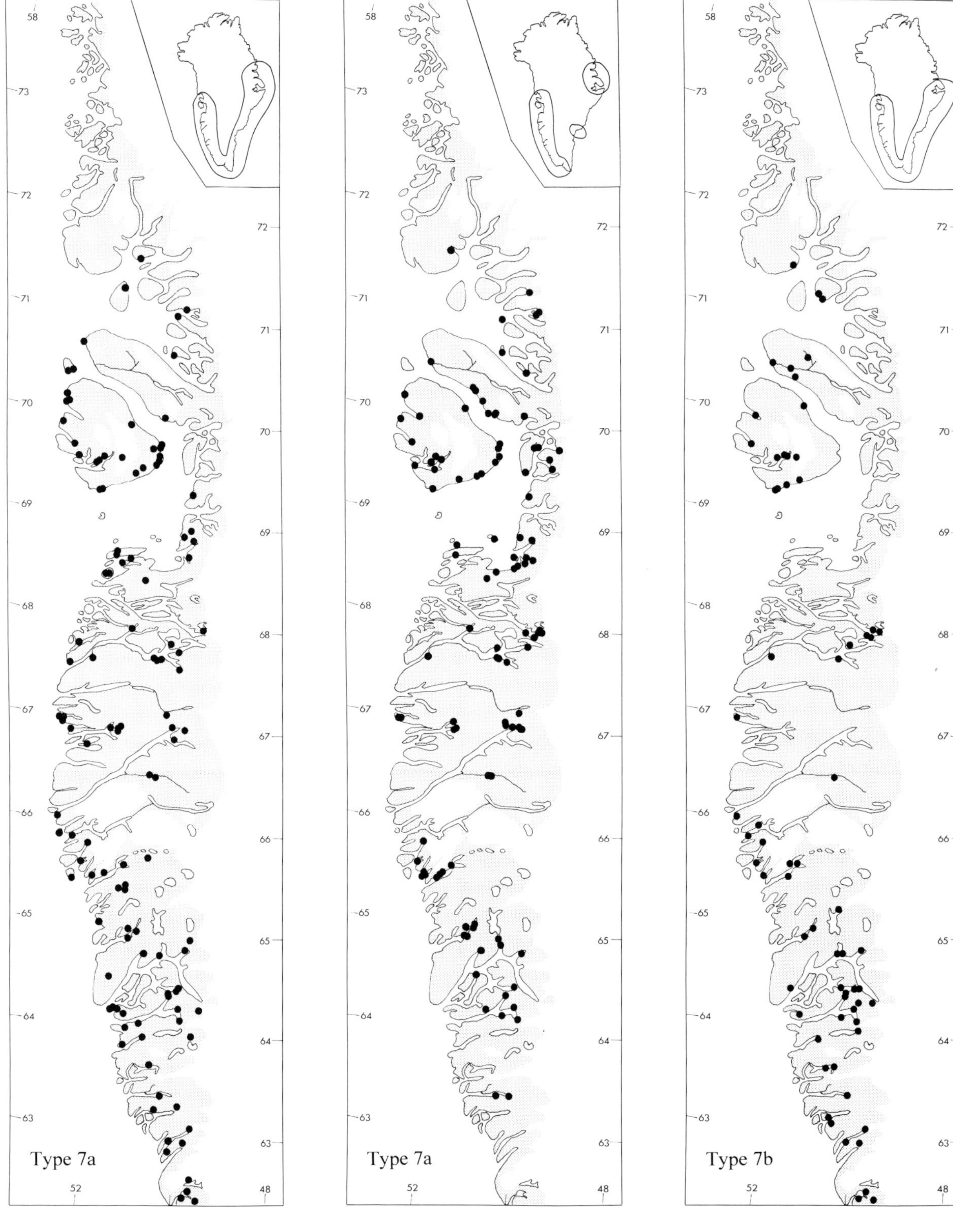

190. Thalictrum alpinum

191. Triglochin palustre

192. Botrychium lunaria

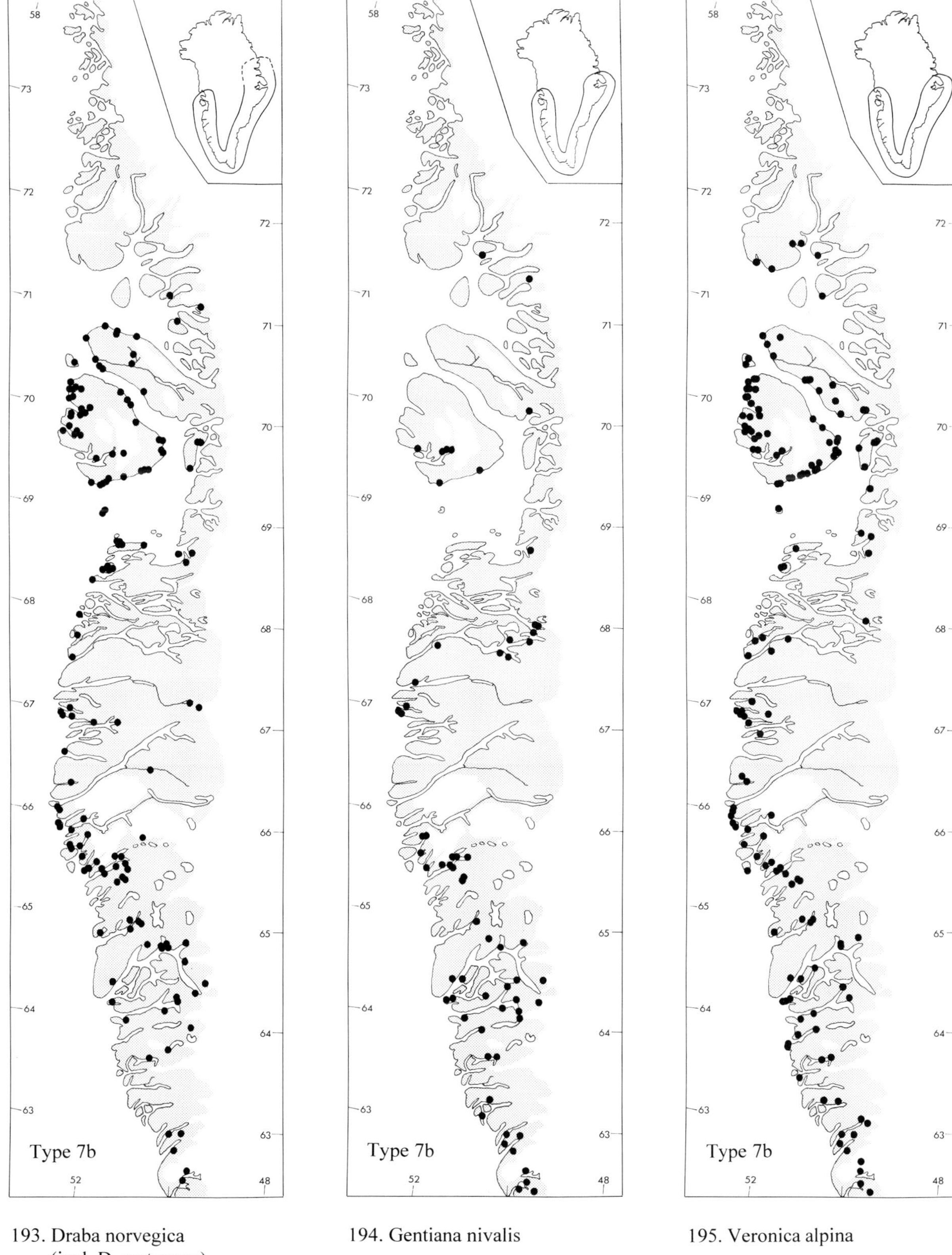

193. Draba norvegica (incl. D. arctogena)

194. Gentiana nivalis

195. Veronica alpina

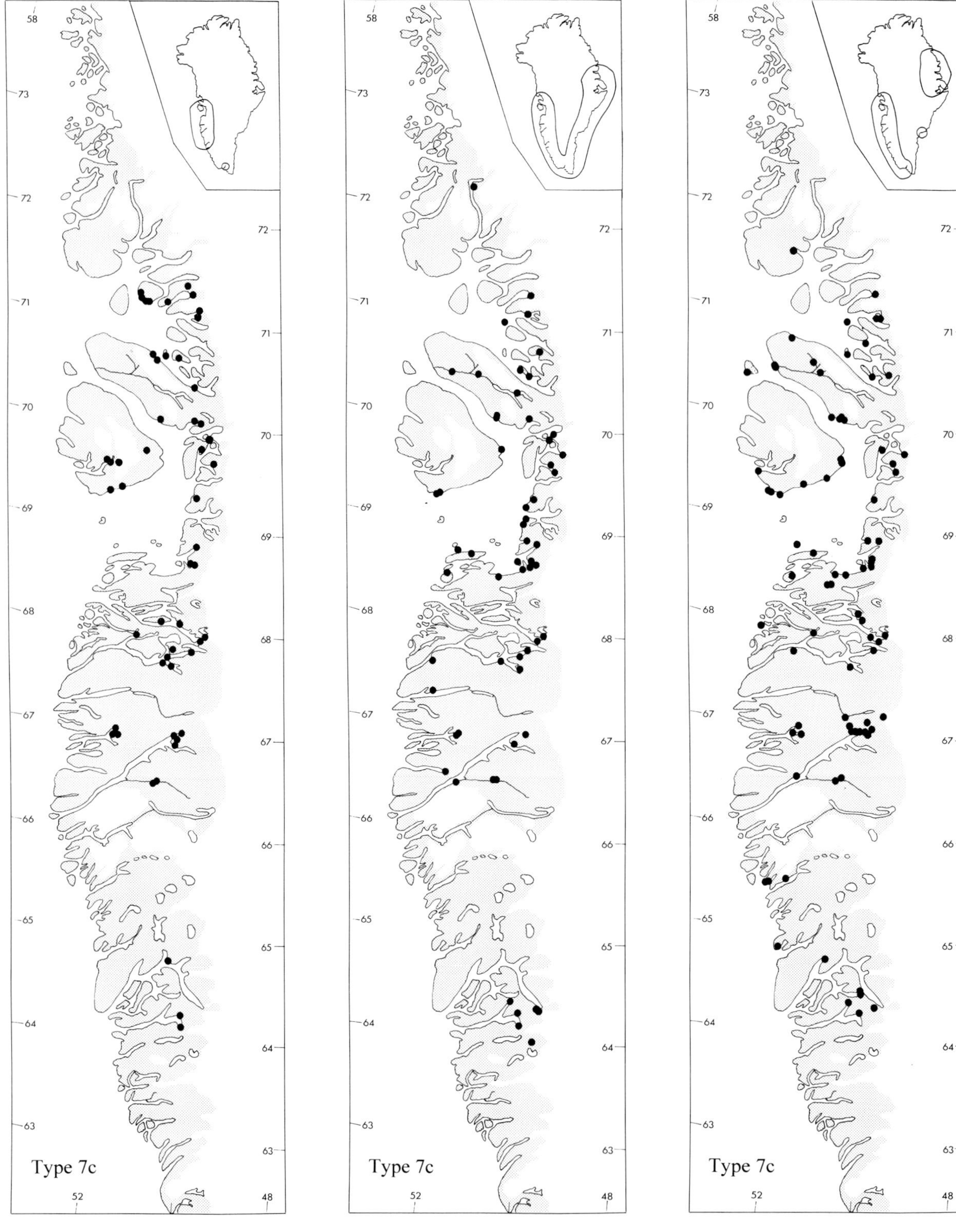

196. Antennaria affinis

197. Carex microglochin

198. Potamogeton filiformis

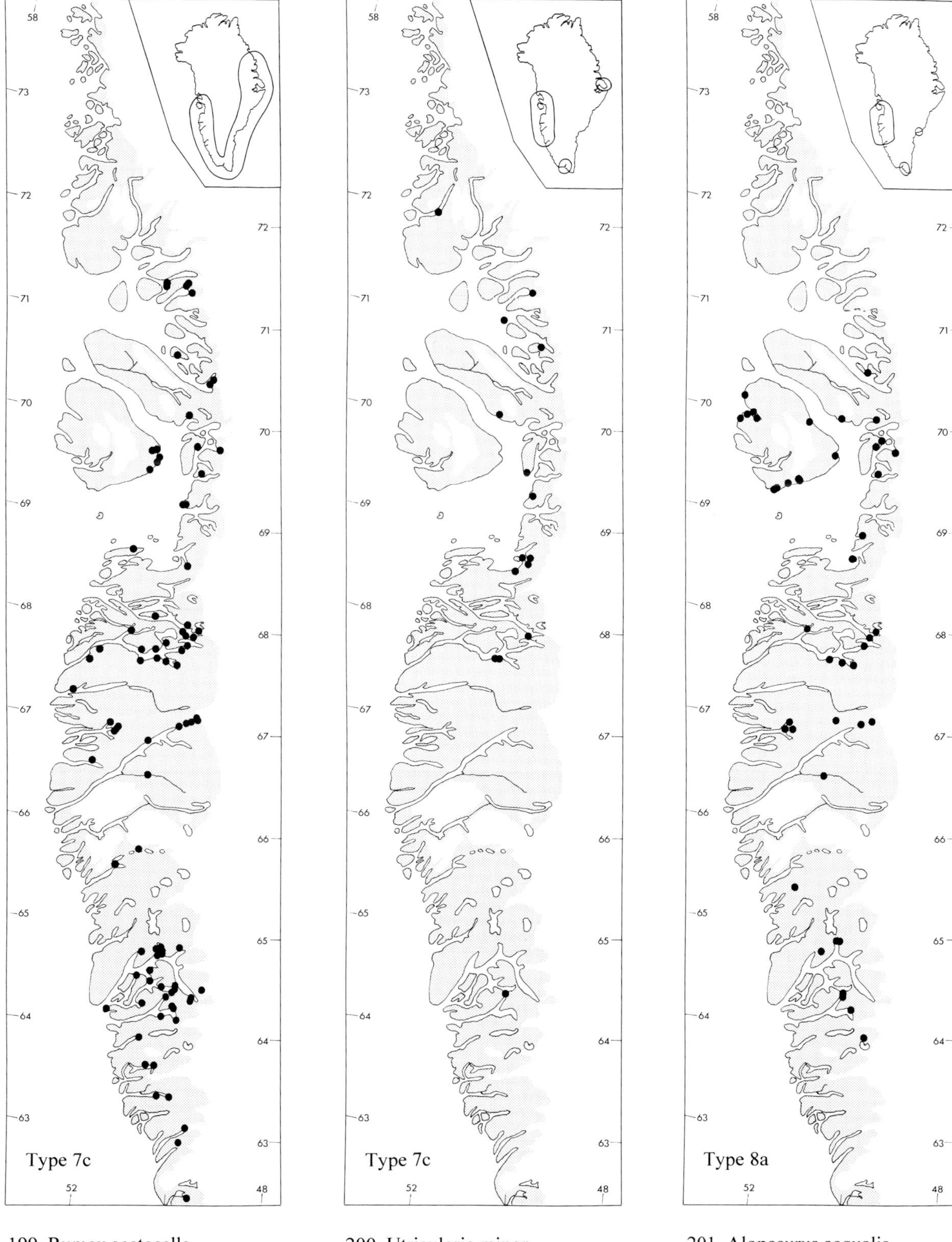

199. Rumex acetosella

200. Utricularia minor

201. Alopecurus aequalis

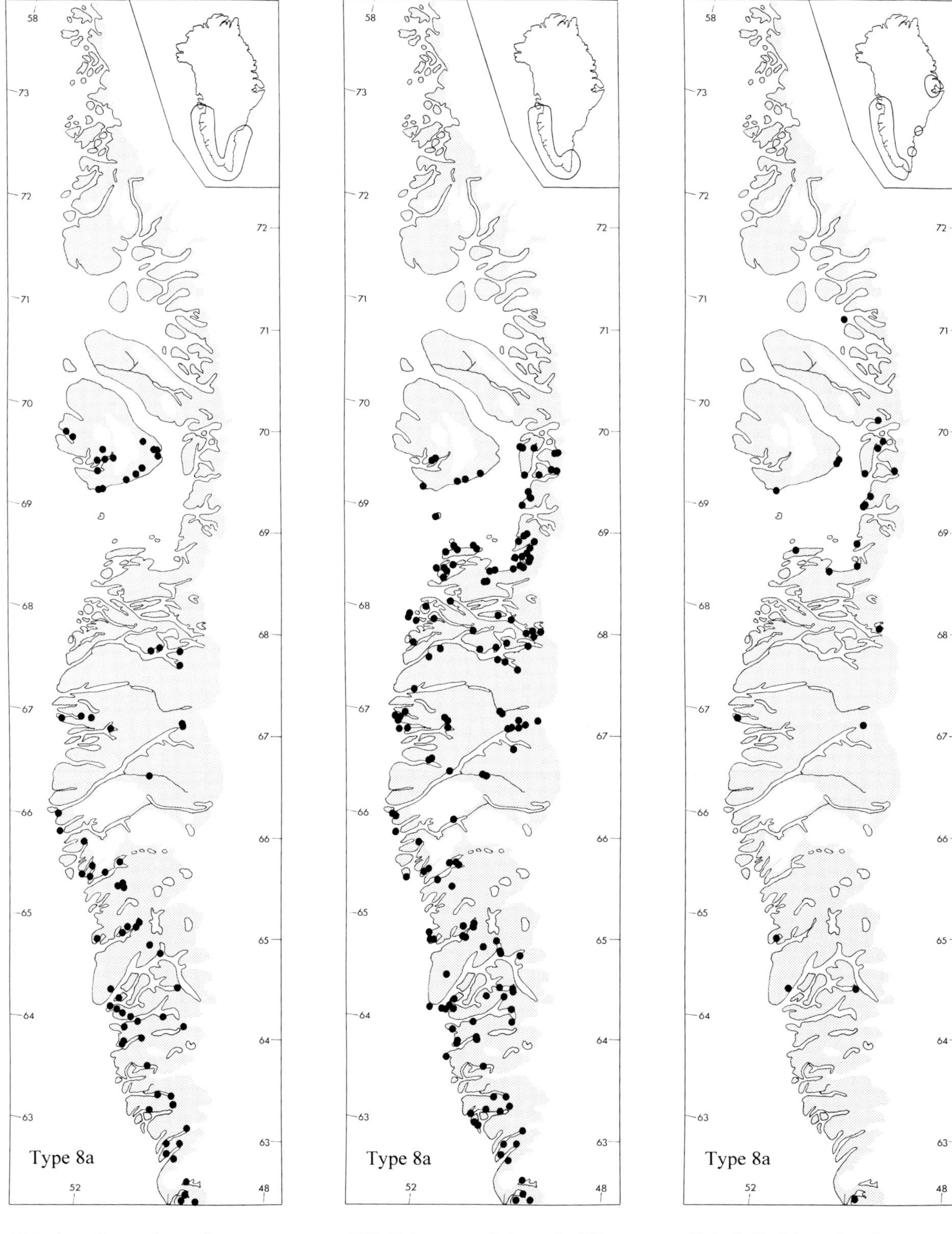

202. Angelica archangelica ssp. norvegica

203. Calamagrostis langsdorffii

204. Callitriche palustris

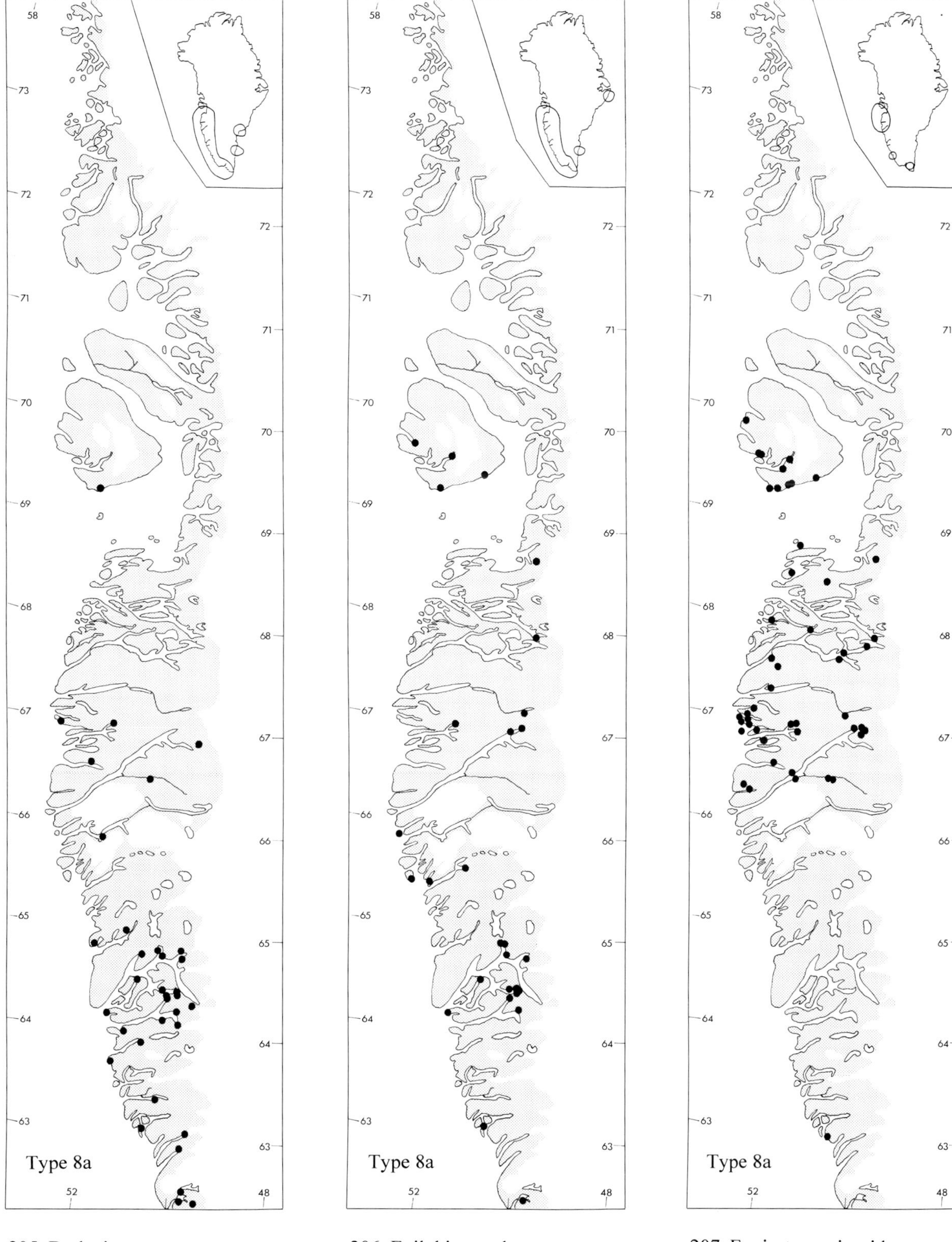

205. Draba incana

206. Epilobium palustre

207. Equisetum scirpoides

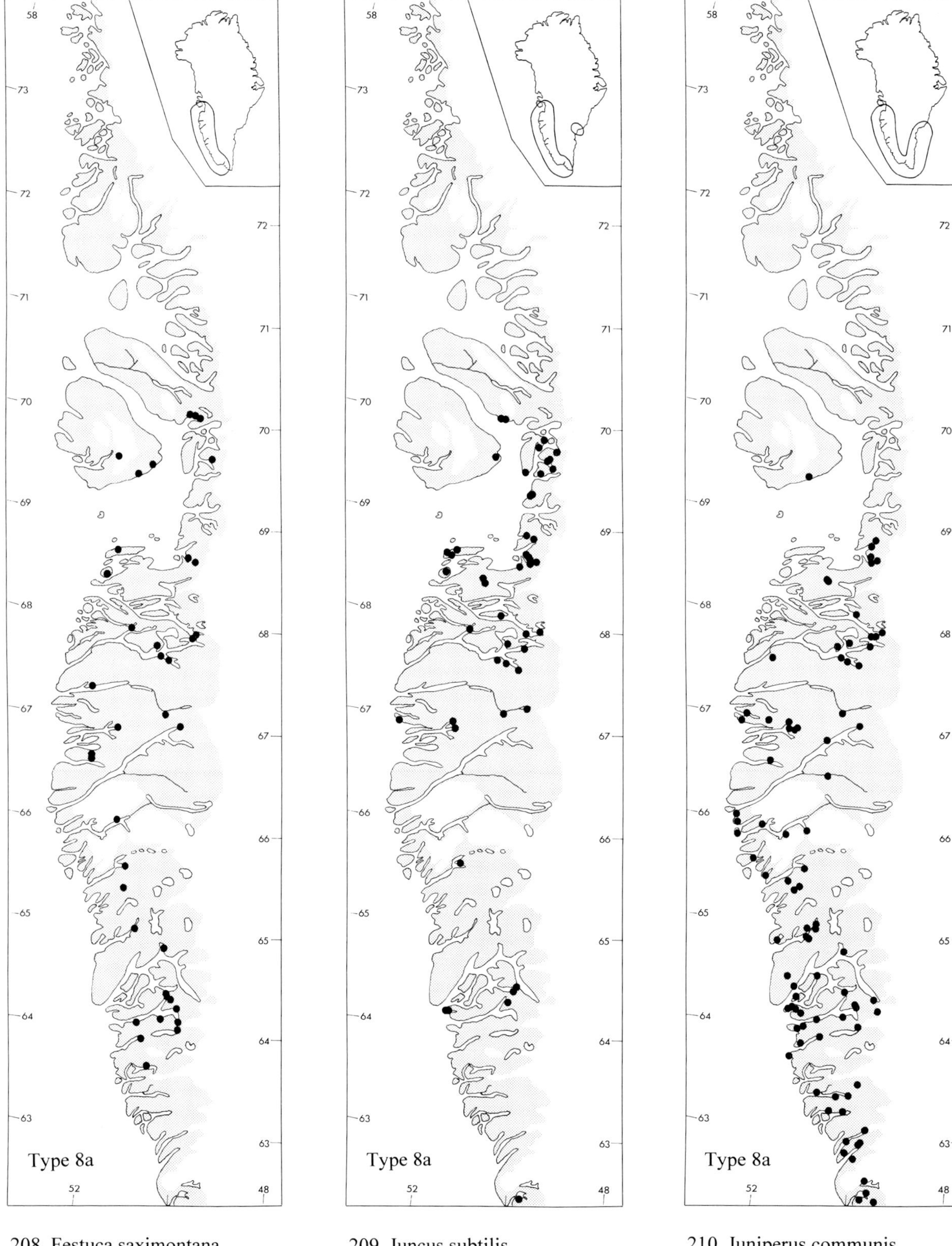

208. Festuca saximontana

209. Juncus subtilis

210. Juniperus communis
ssp. alpina

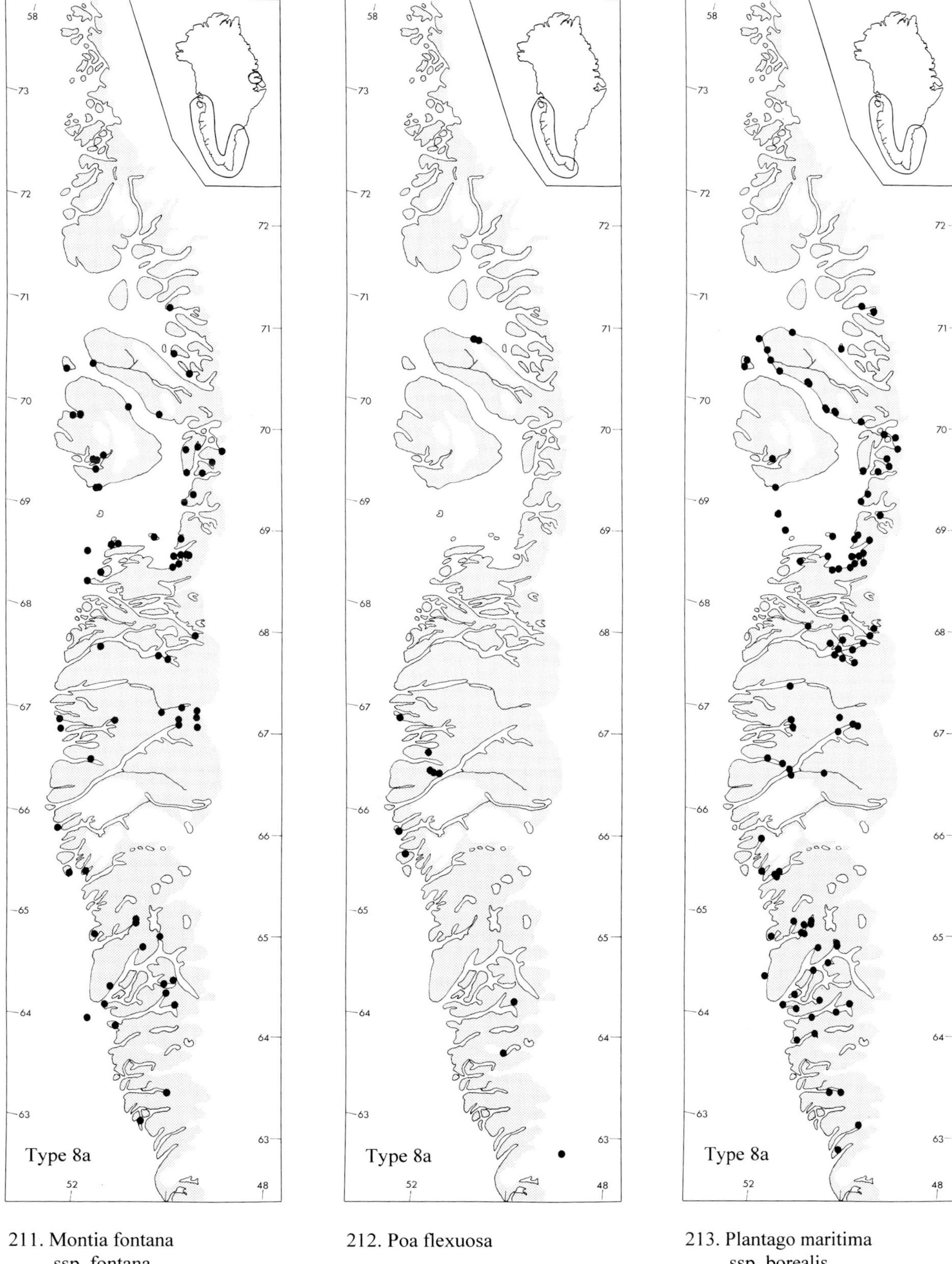

211. Montia fontana ssp. fontana

212. Poa flexuosa

213. Plantago maritima ssp. borealis

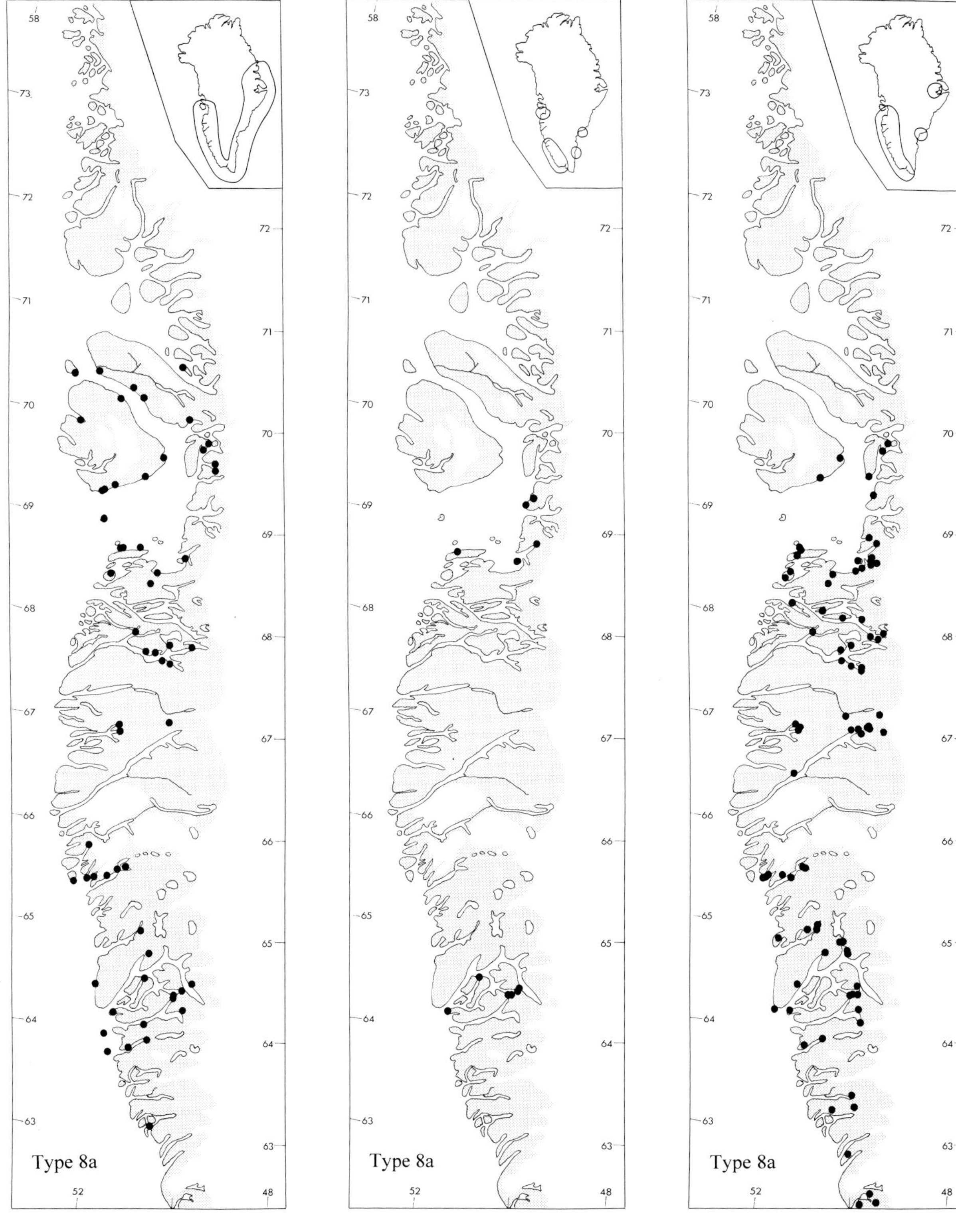

214. Puccinellia coarctata

215. Sparganium angustifolium

216. Sparganium hyperboreum

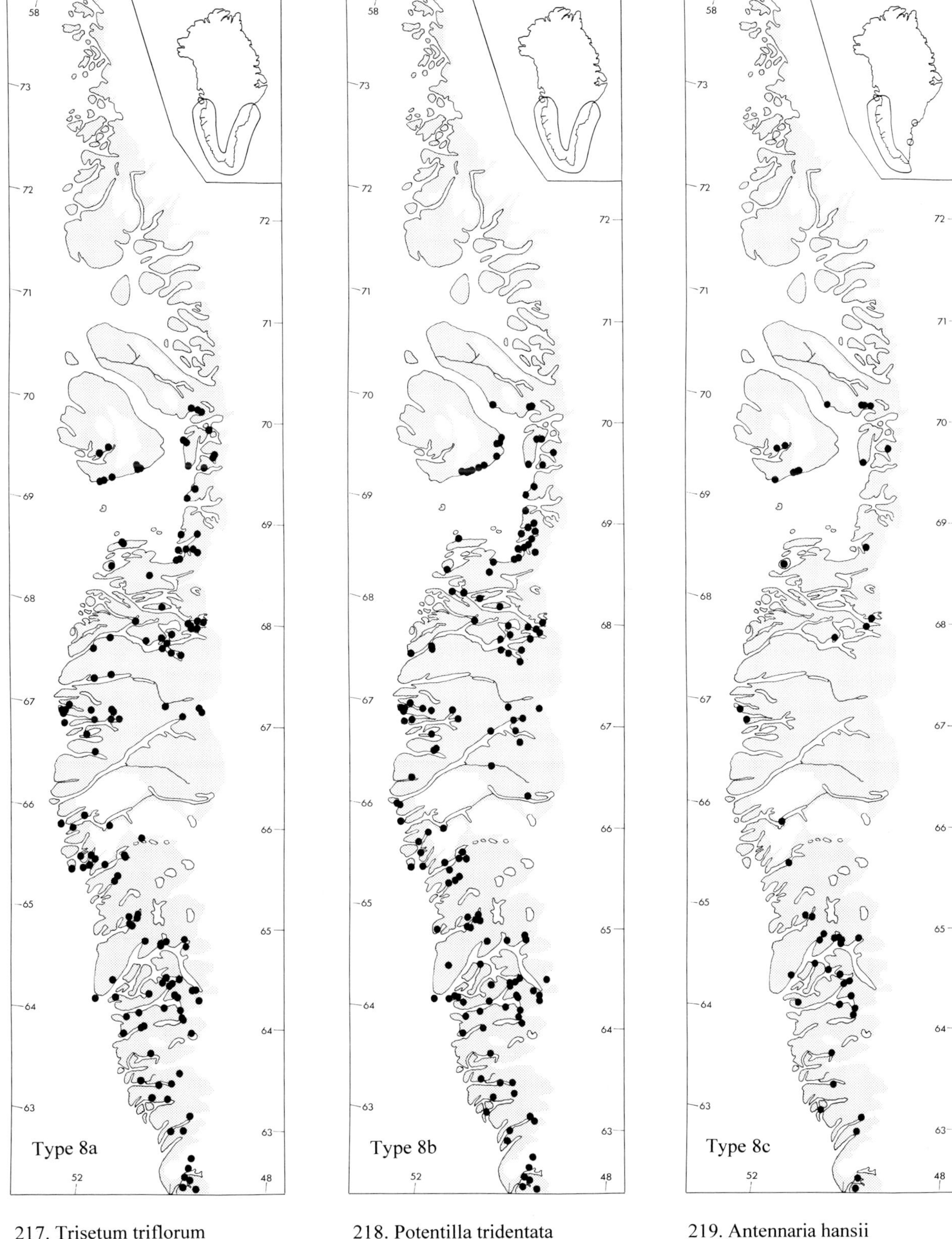

217. Trisetum triflorum

218. Potentilla tridentata

219. Antennaria hansii

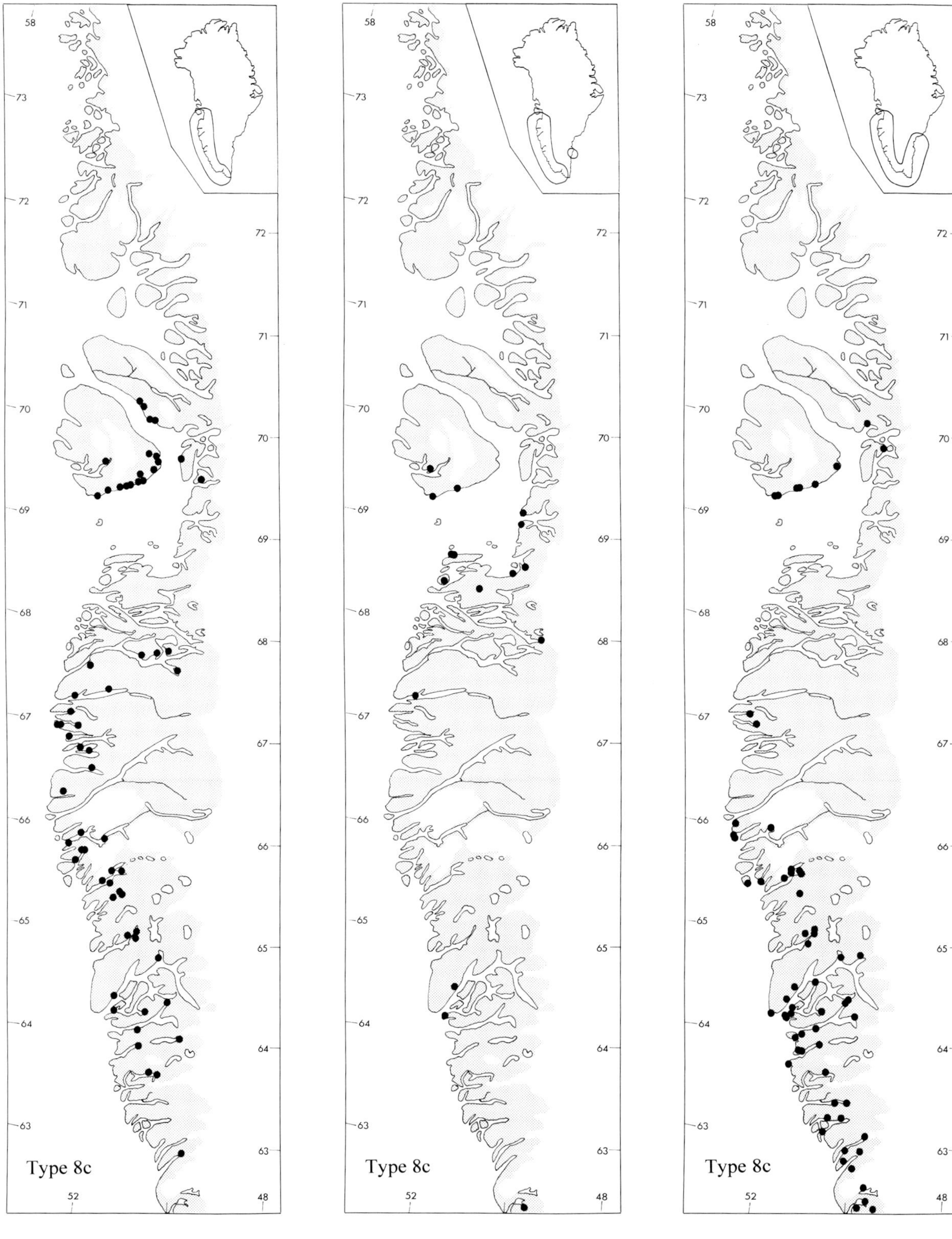

220. Antennaria intermedia

221. Callitriche anceps

222. Carex brunnescens

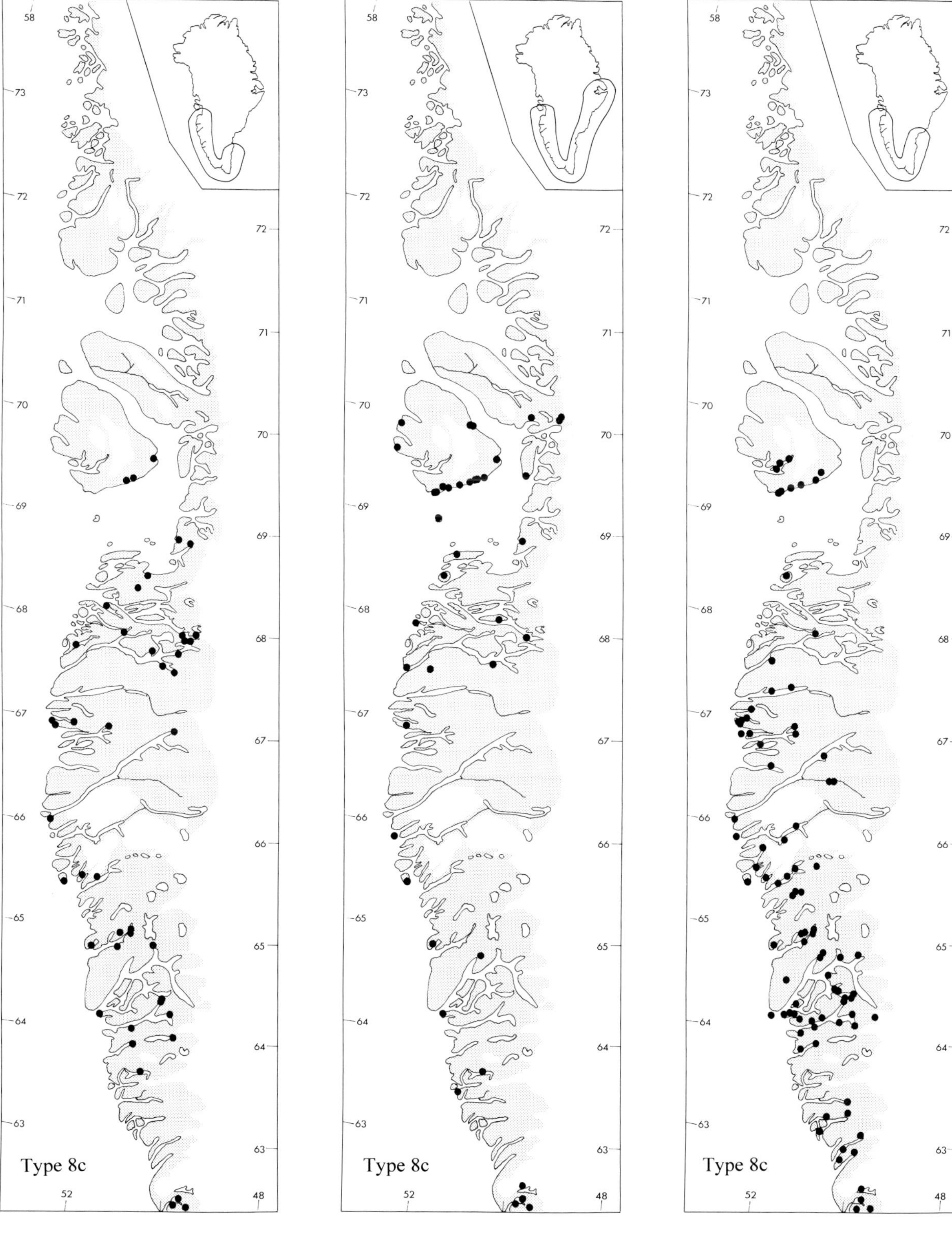

223. Carex canescens

224. Carex rufina

225. Chamaenerion angustifolium

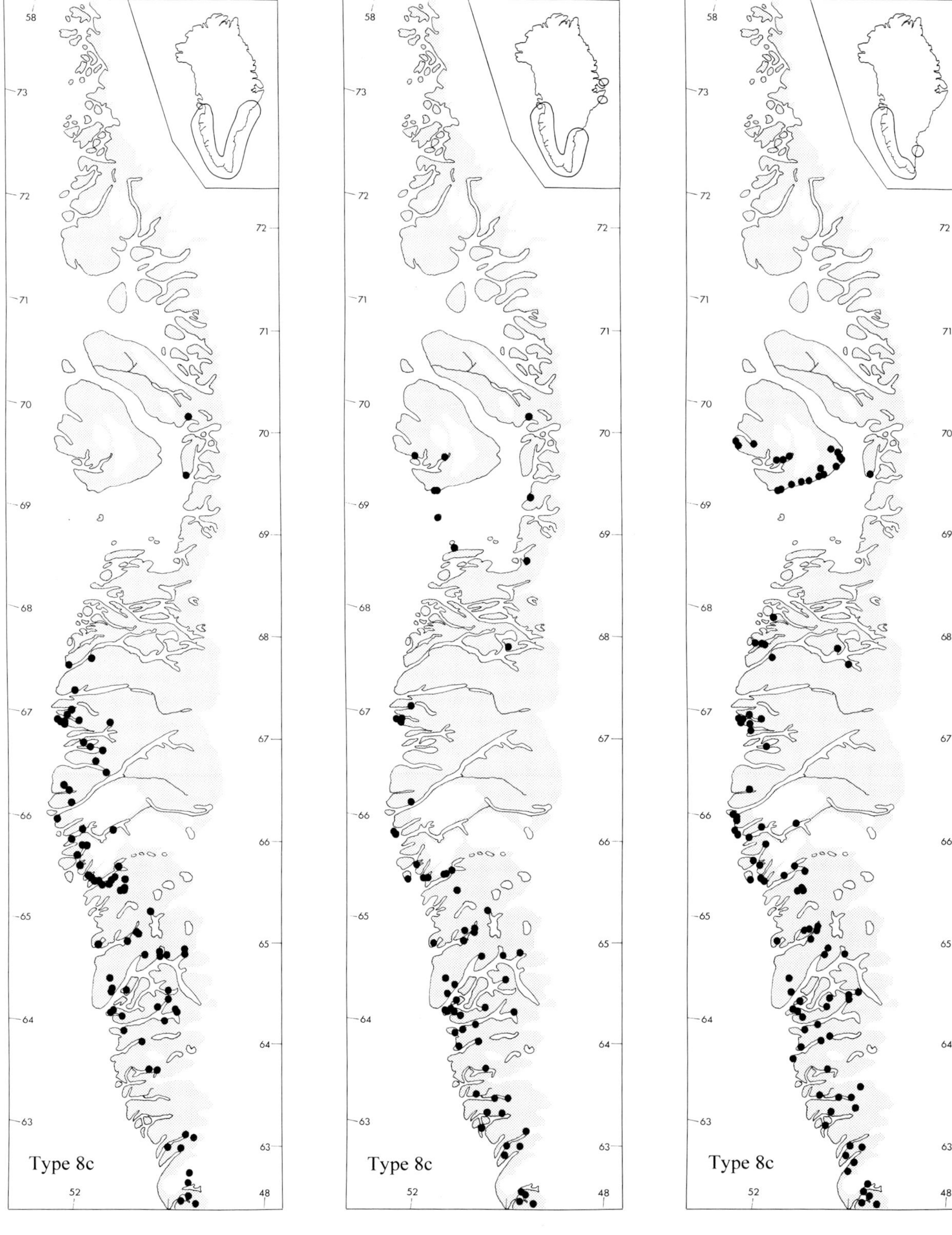

226. Erigeron uniflorus

227. Gymnocarpium dryopteris

228. Luzula parviflora

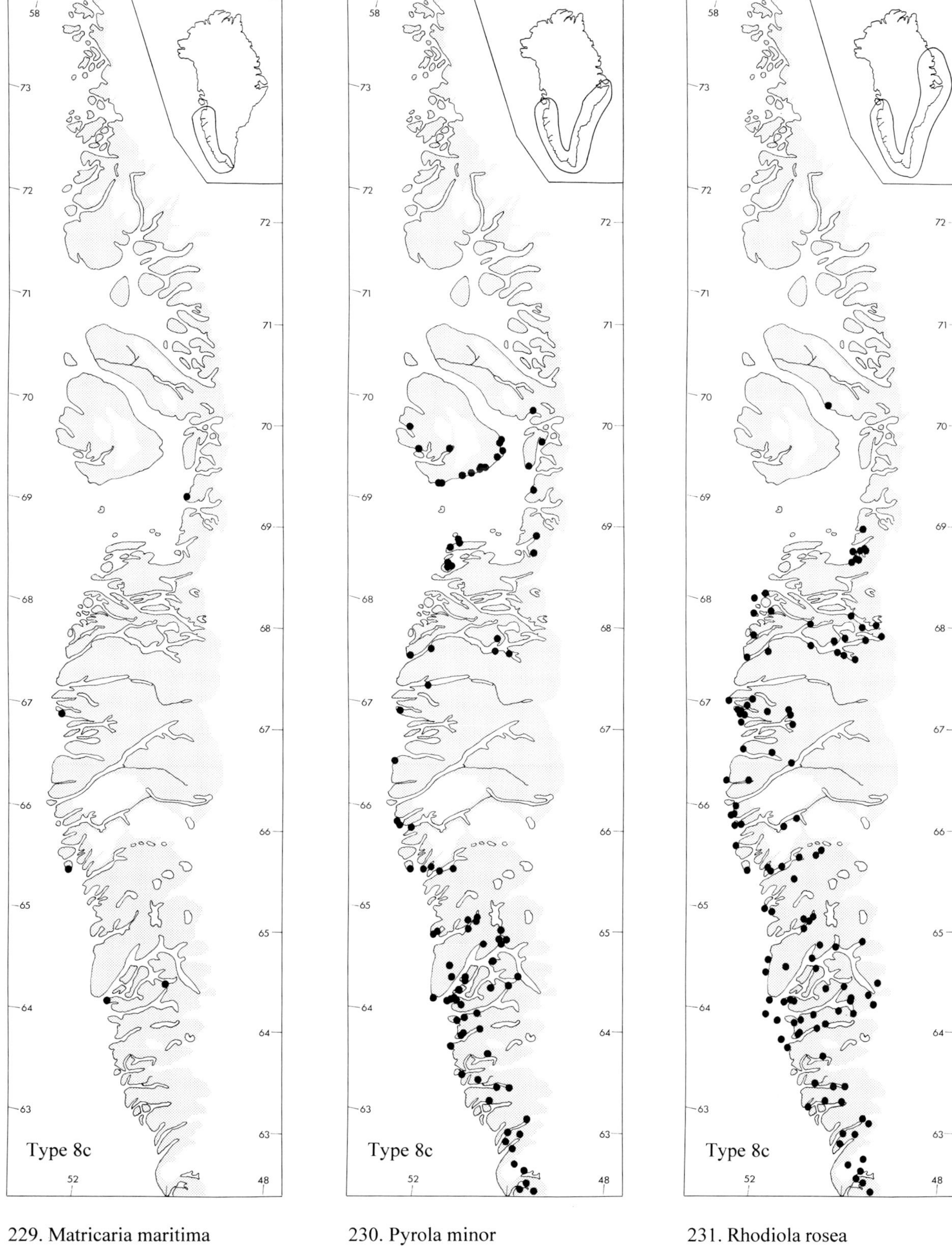

229. Matricaria maritima ssp. borealis

230. Pyrola minor

231. Rhodiola rosea

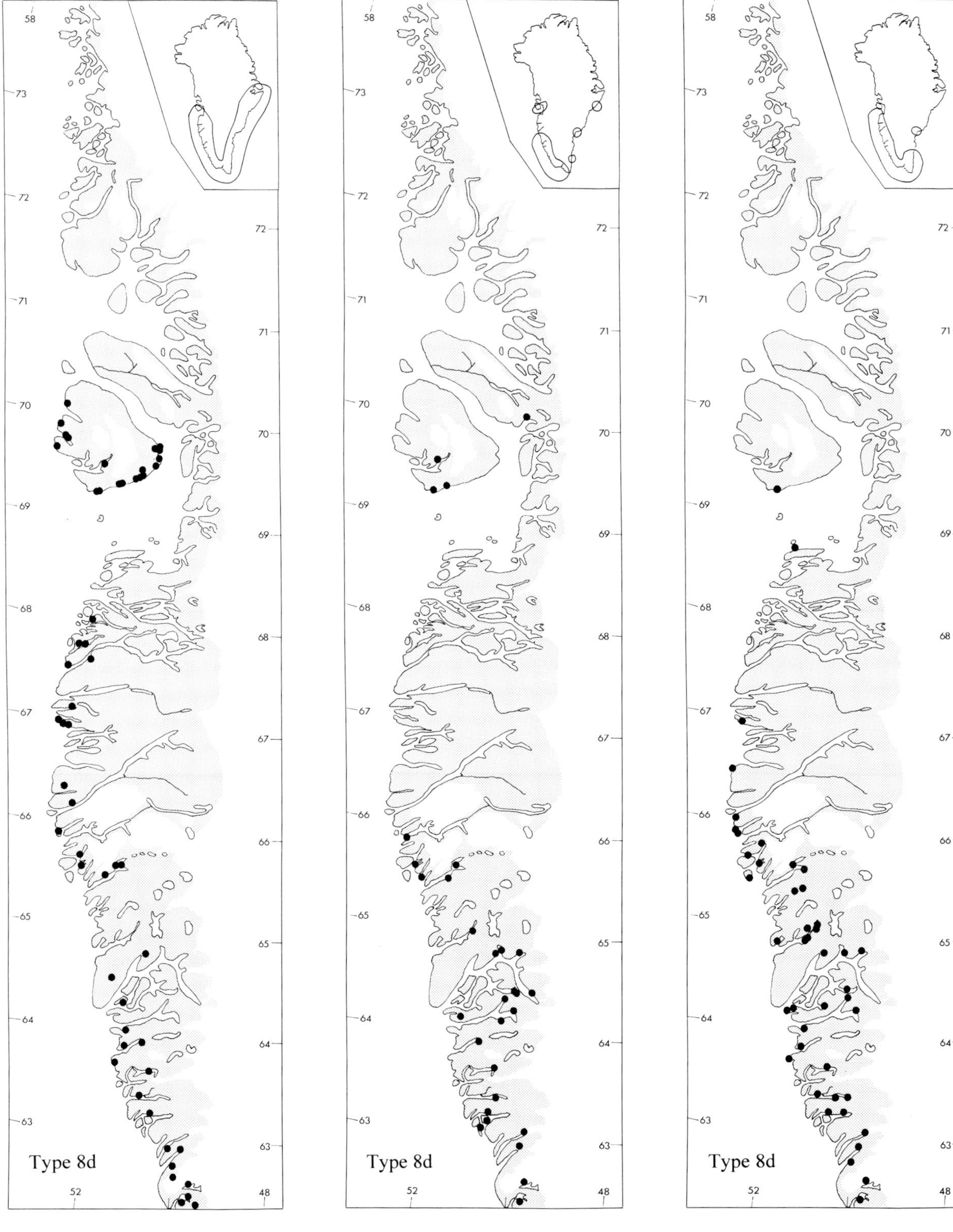

232. Alchemilla glomerulans

233. Botrychium lanceolatum

234. Carex deflexa

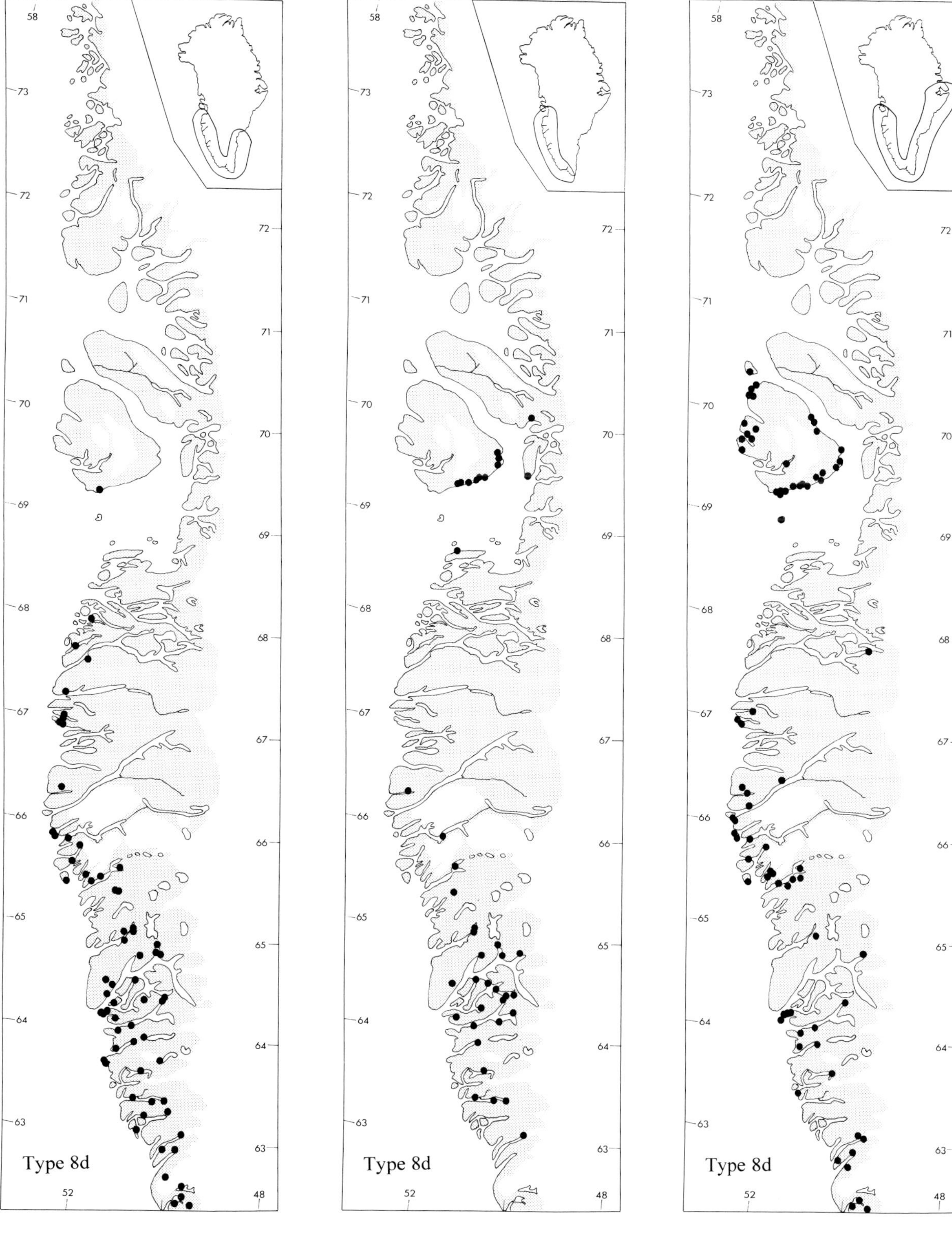

235. Coptis trifolia

236. Diphasiastrum complanatum

237. Epilobium anagallidifolium

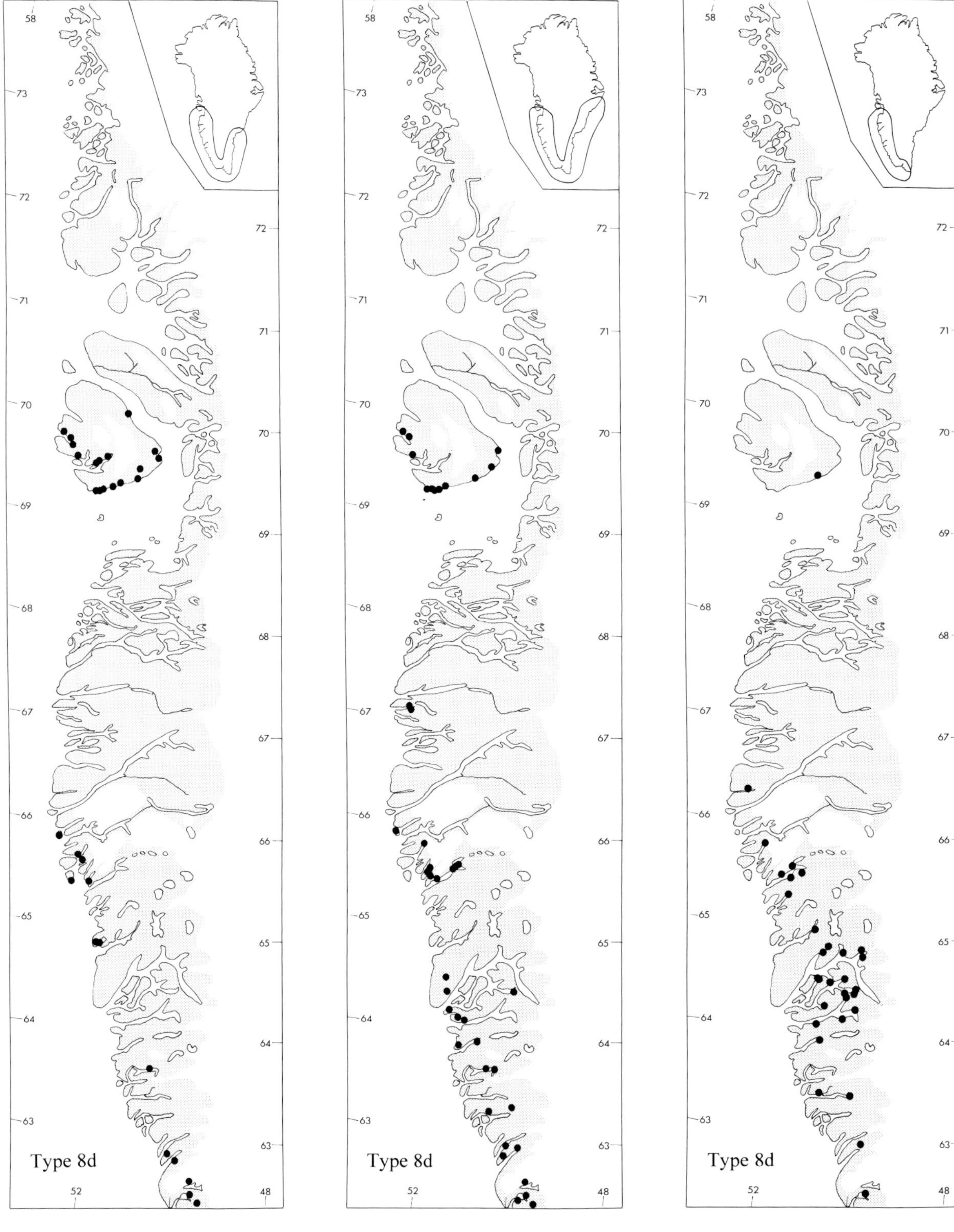

238. Epilobium hornemannii

239. Epilobium lactiflorum

240. Equisetum sylvaticum

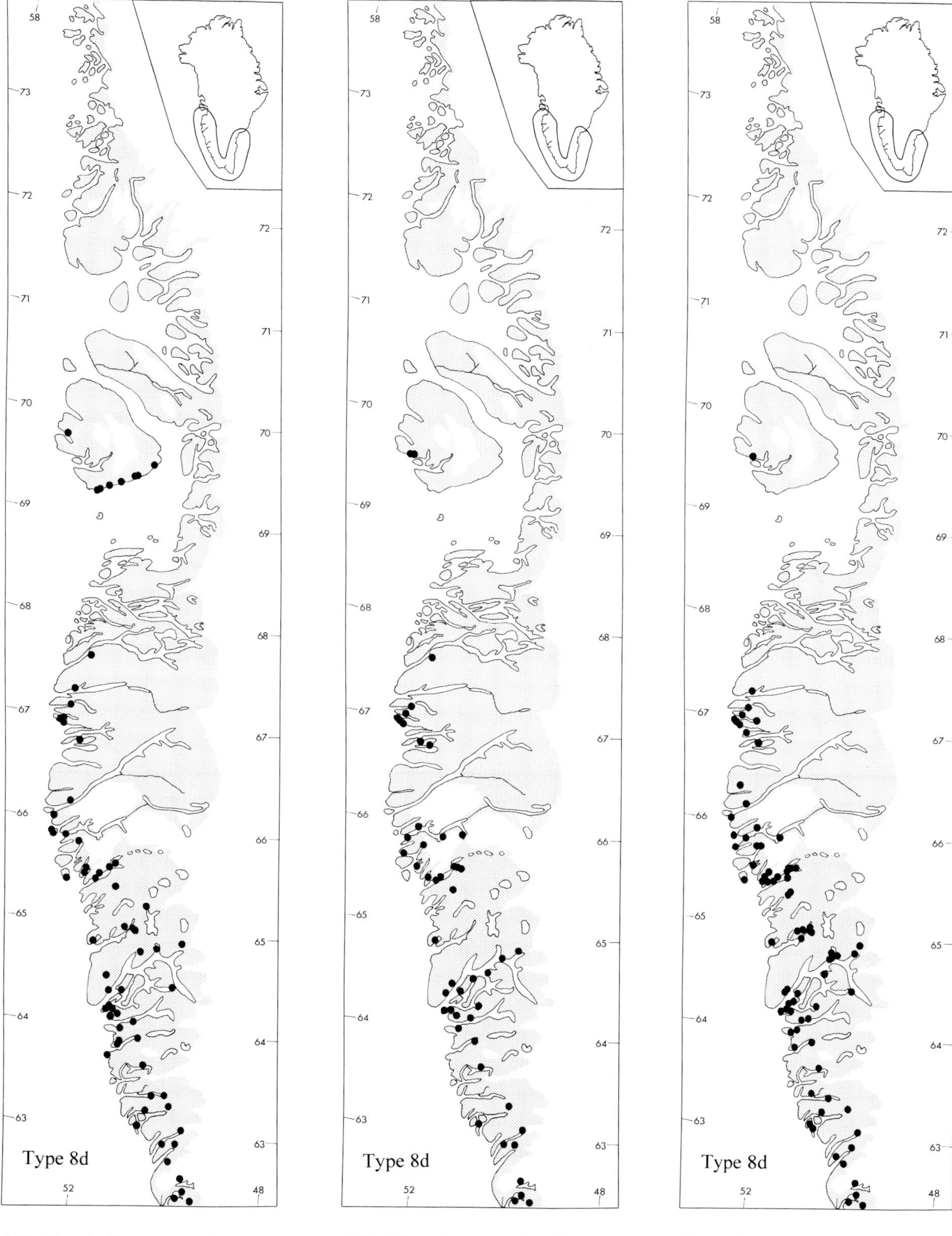

241. Gnaphalium norvegicum

242. Hieracium groenlandicum

243. Hieracium hyparcticum

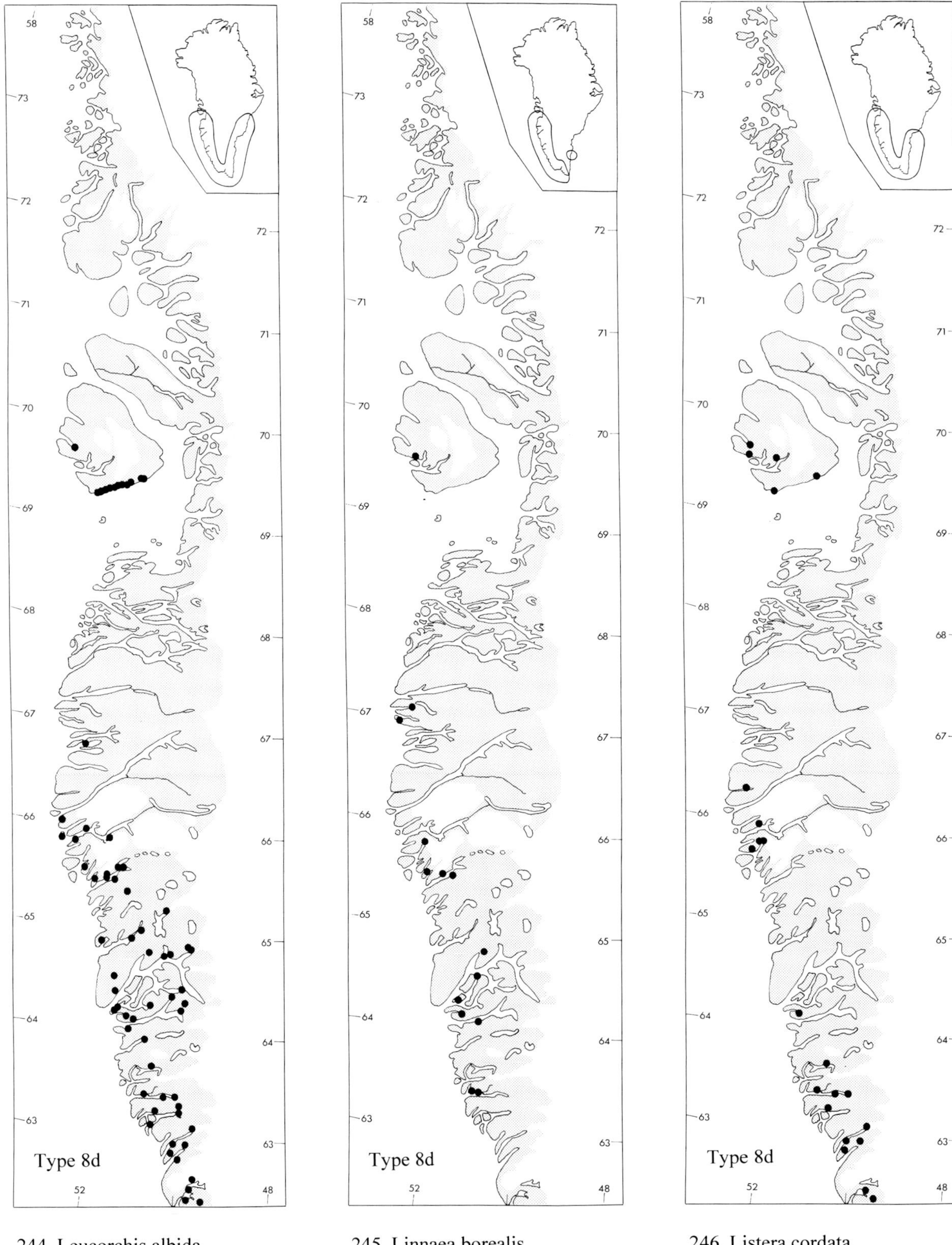

244. Leucorchis albida ssp. straminea

245. Linnaea borealis ssp. americana

246. Listera cordata

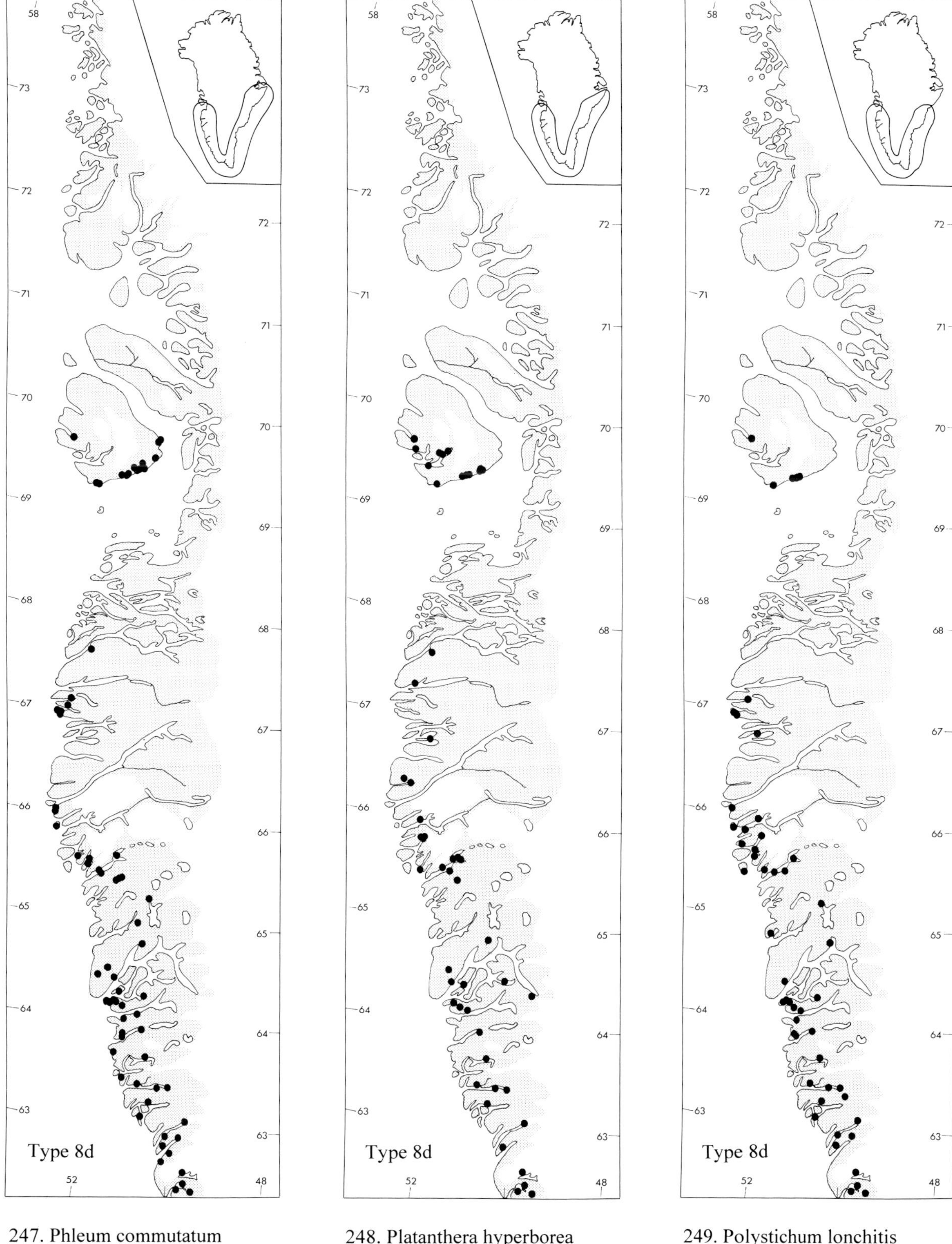

247. Phleum commutatum

248. Platanthera hyperborea

249. Polystichum lonchitis

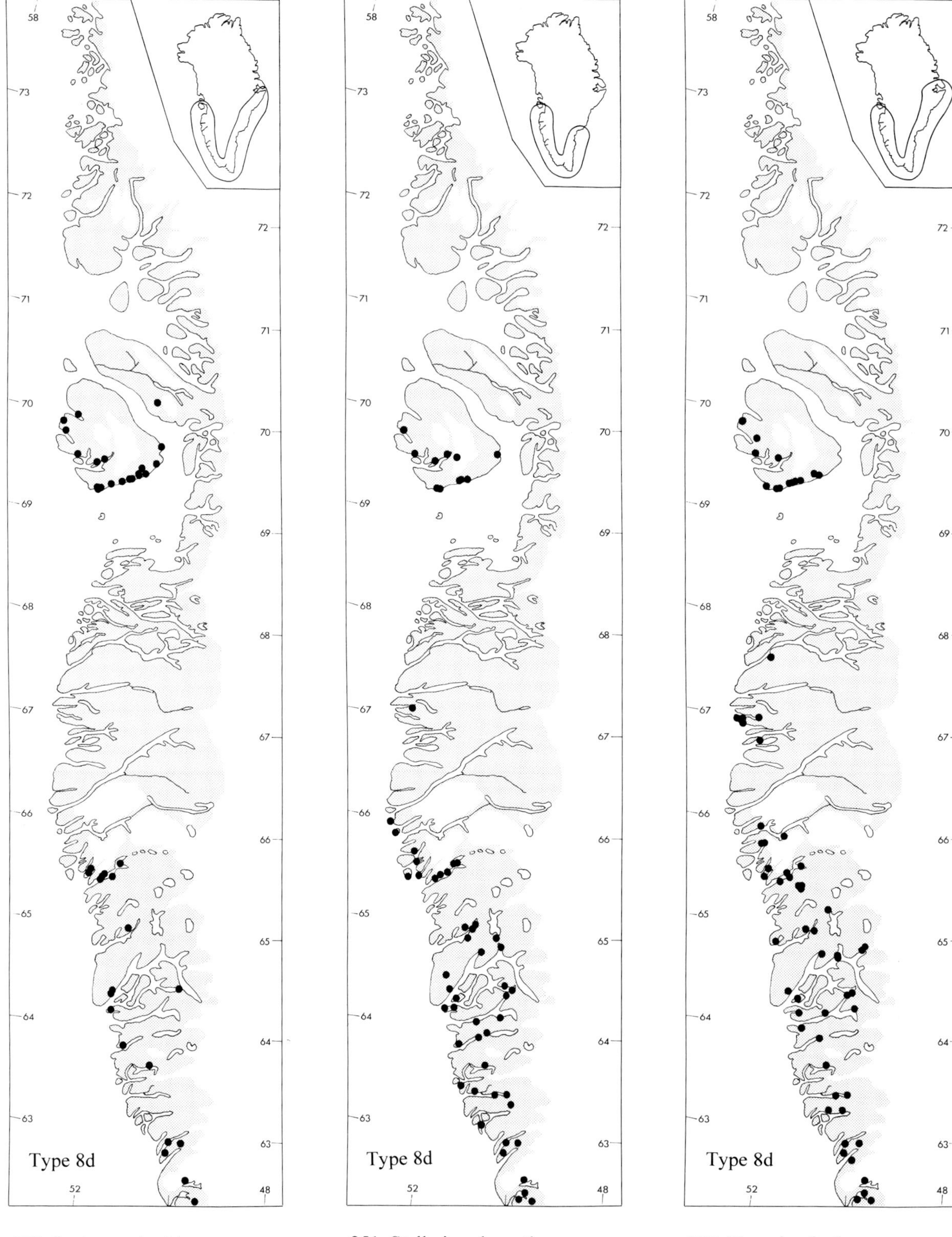

250. Sagina saginoides

251. Stellaria calycantha

252. Veronica fruticans

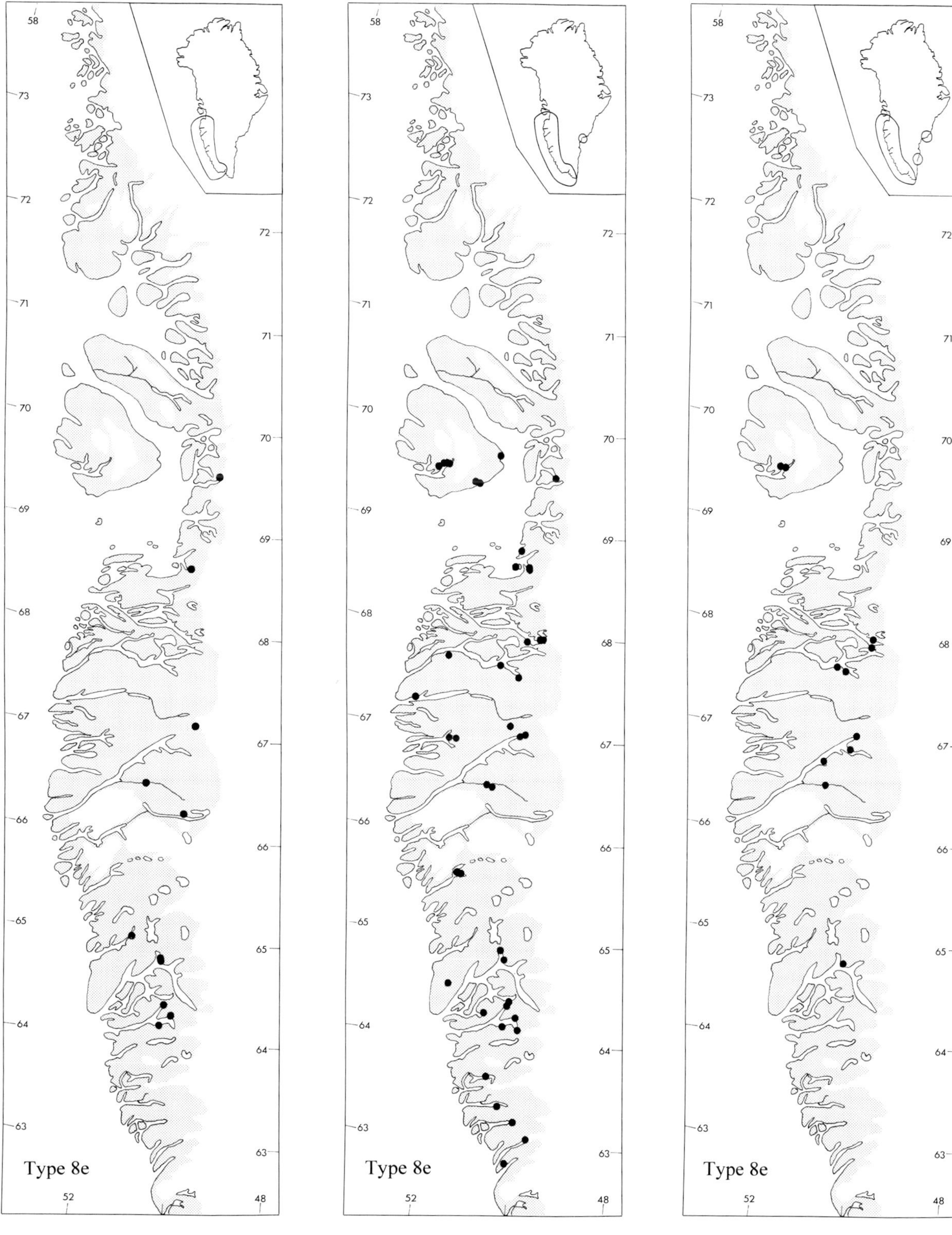

253. Calamagrostis poluninii

254. Corallorrhiza trifida

255. Gentiana aurea

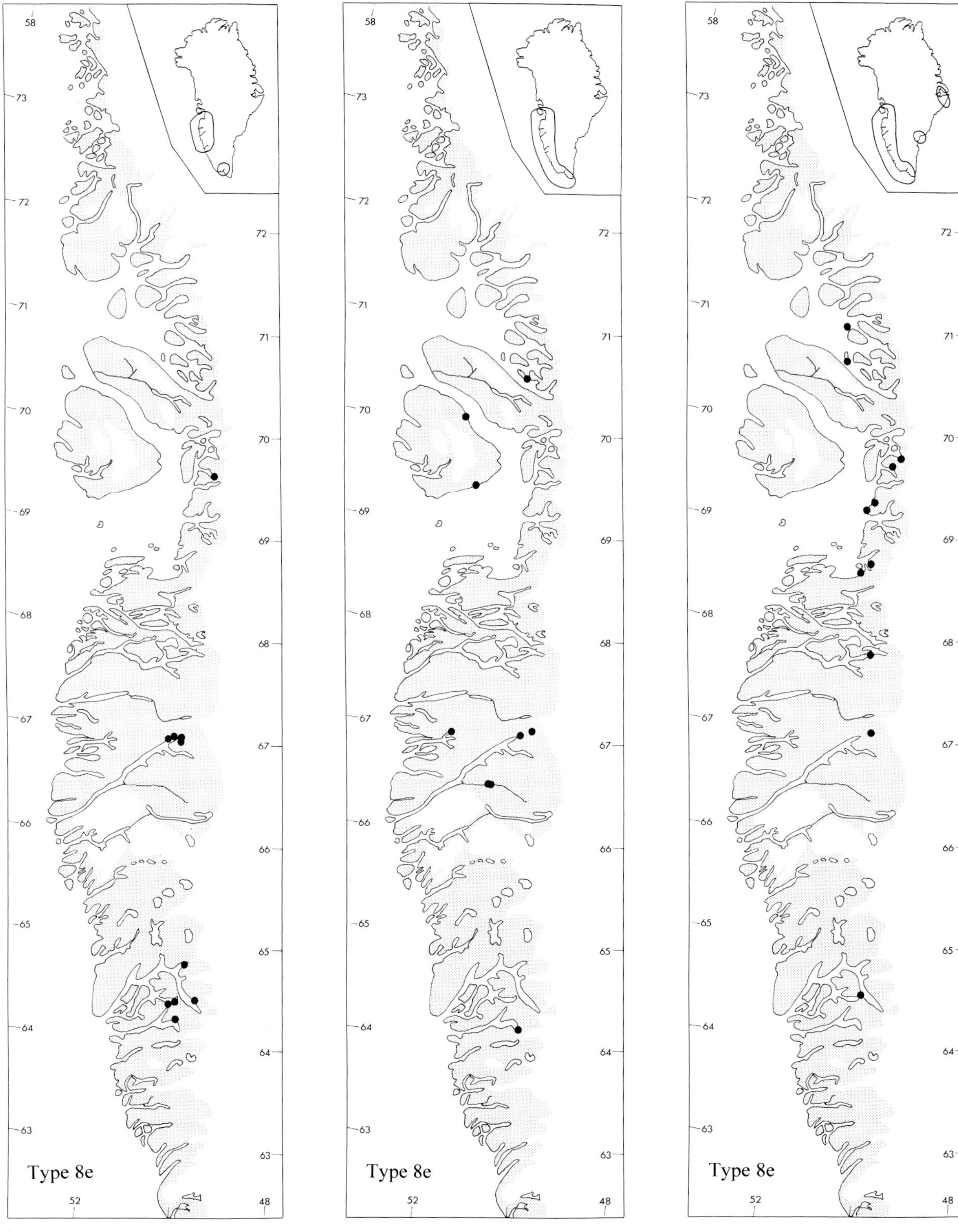

256. Juncus alpinus ssp. nodulosus

257. Leymus arenarius

258. Limosella aquatica

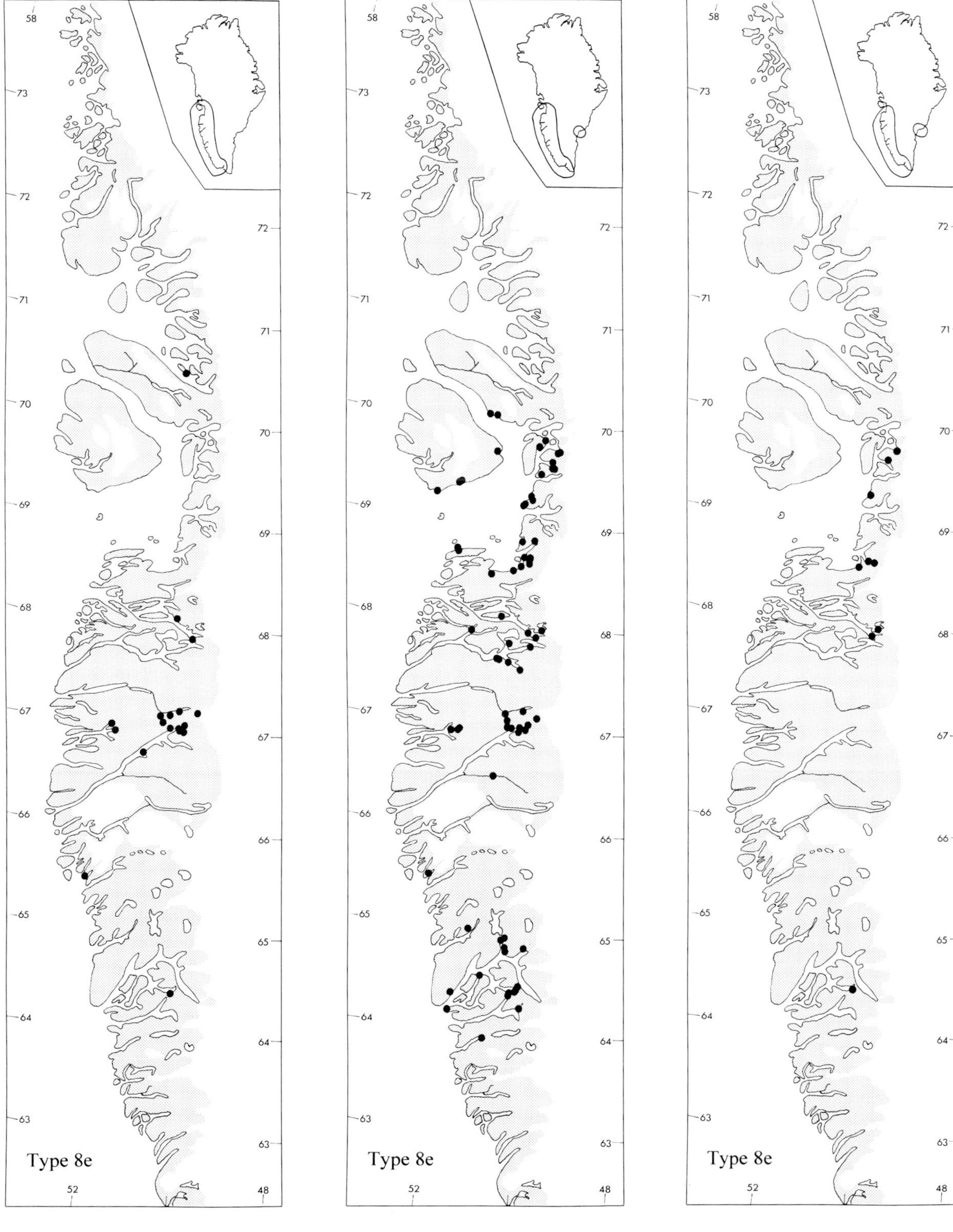

259. Myriophyllum spicatum ssp. exalbescens

260. Ranunculus reptans

261. Subularia aquatica

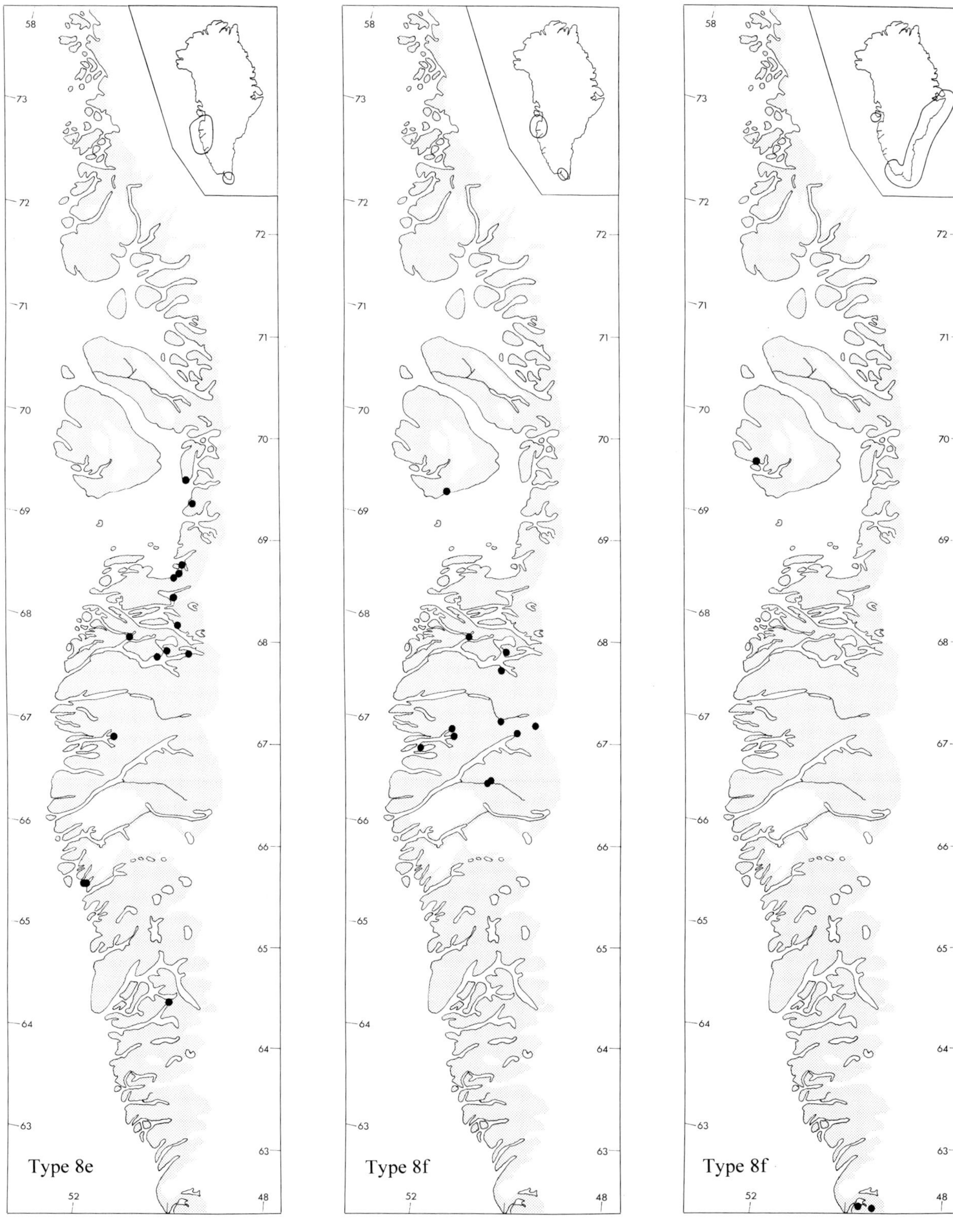

262. Utricularia intermedia

263. Elymus violaceus

264. Hieracium alpinum

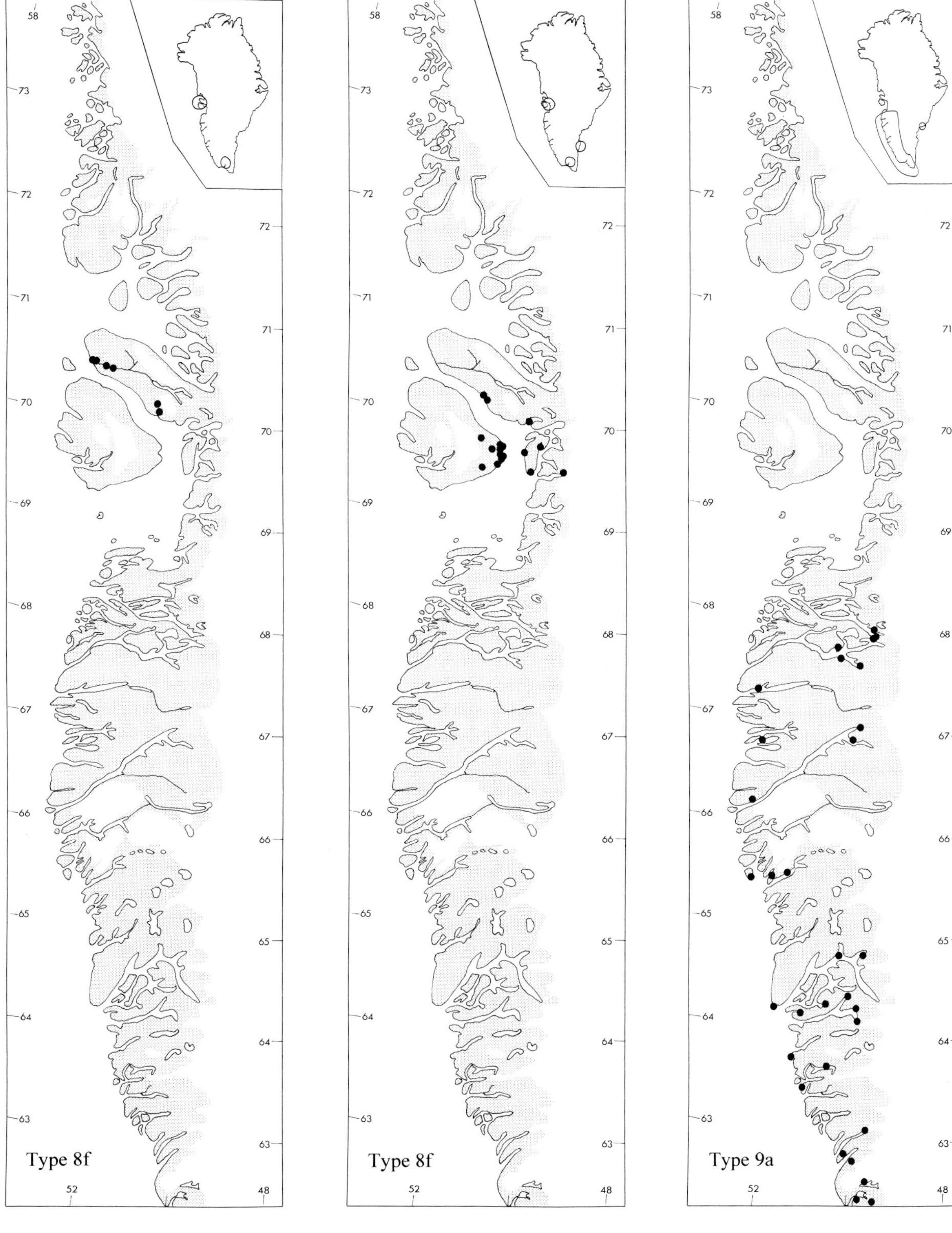

265. Parnassia kotzebuei

266. Potentilla ranunculus

267. Agrostis hyperborea

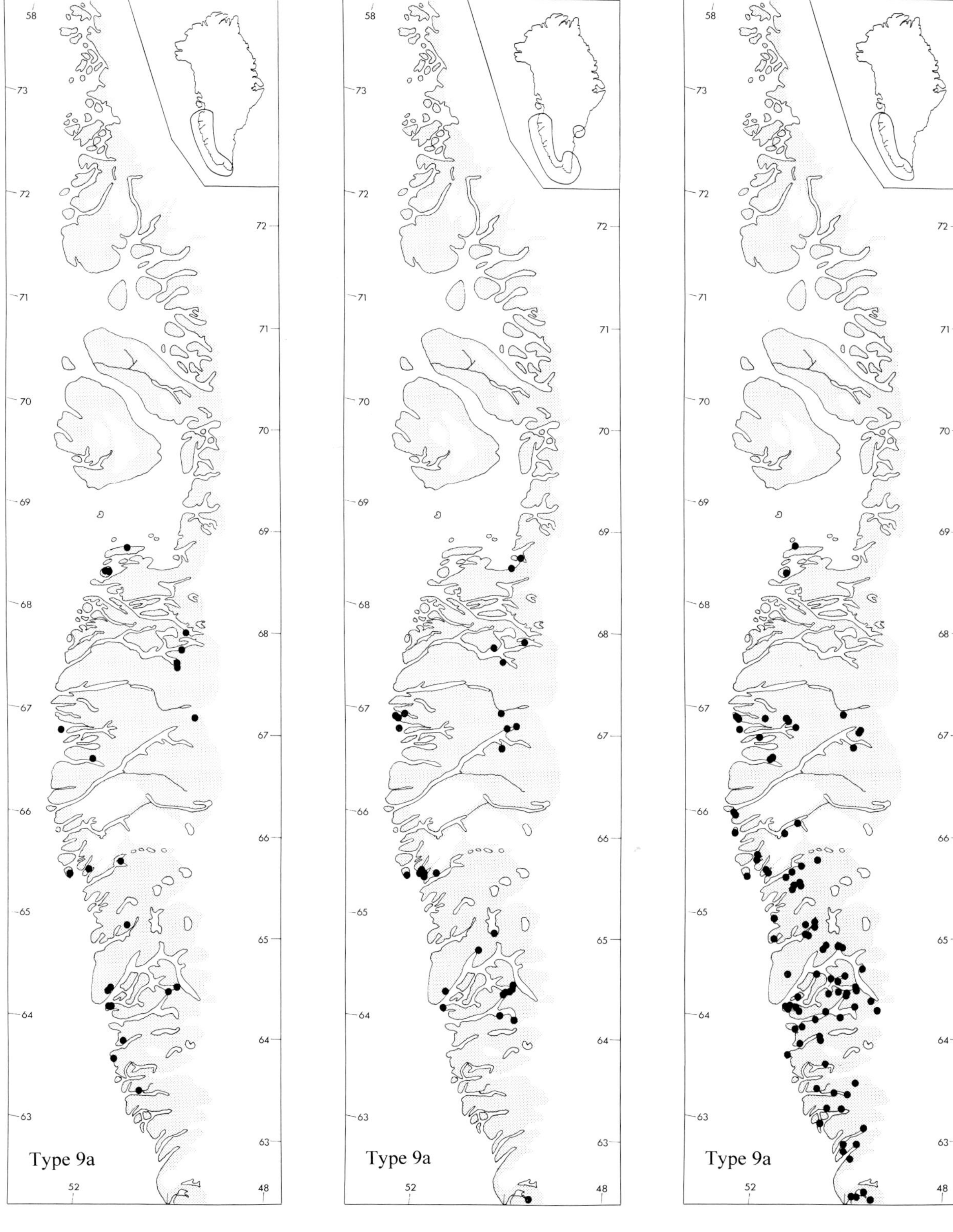

268. Callitriche hamulata

269. Comarum palustre

270. Ledum groenlandicum

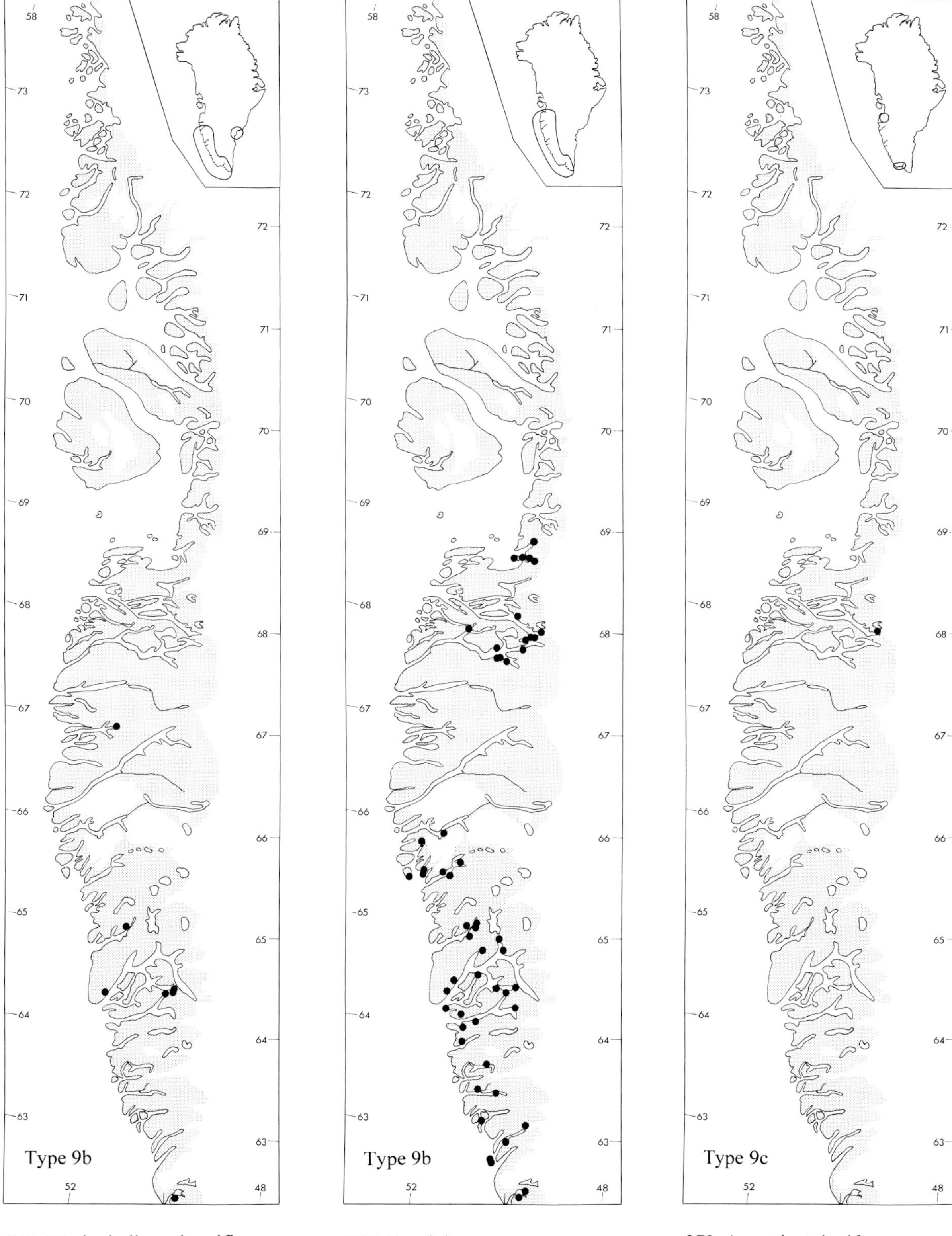

271. Myriophyllum alterniflorum

272. Vaccinium oxycoccus ssp. microphyllum

273. Agrostis stolonifera

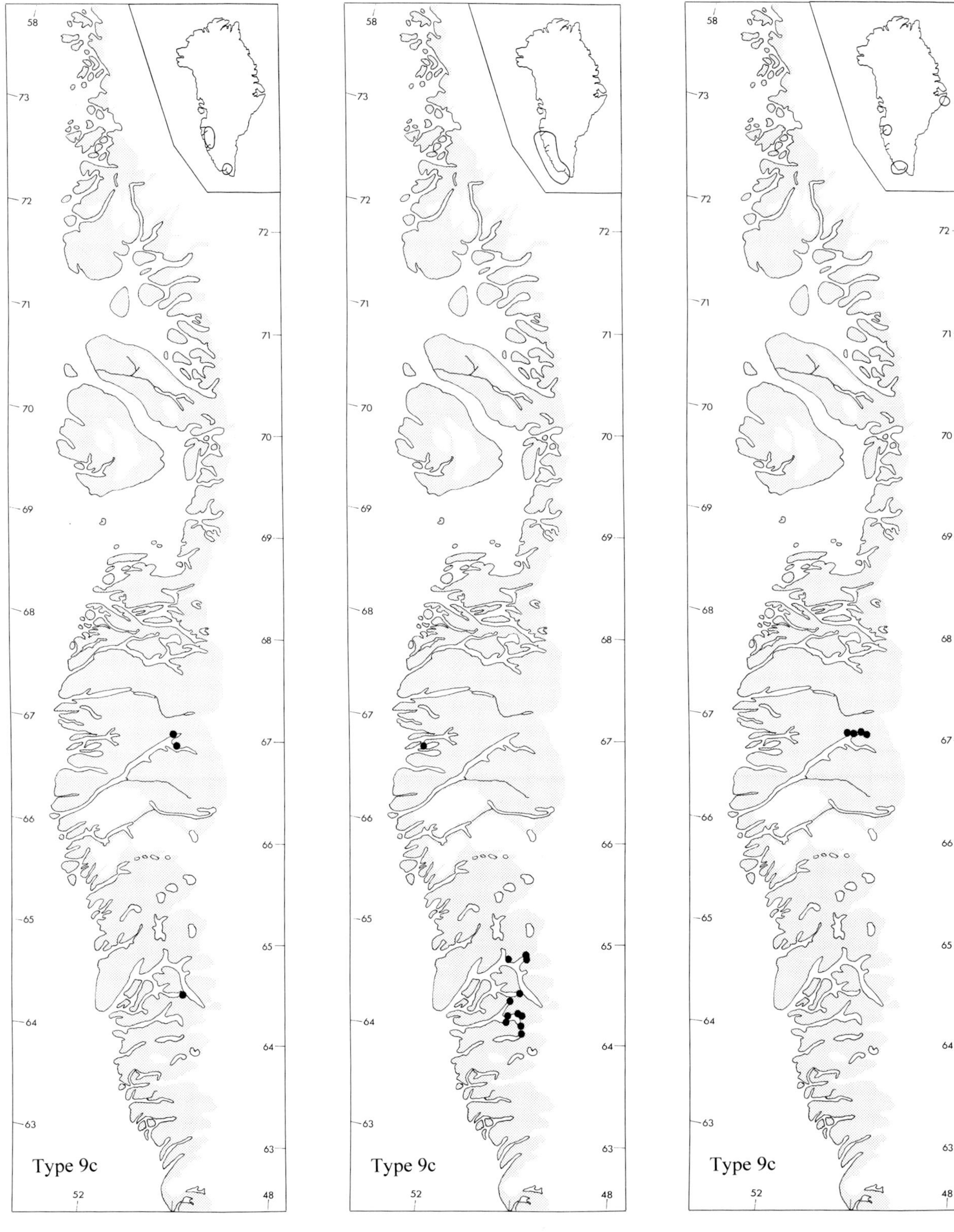

274. Amerorchis rotundifolia

275. Carex praticola

276. Eleocharis quinqueflora

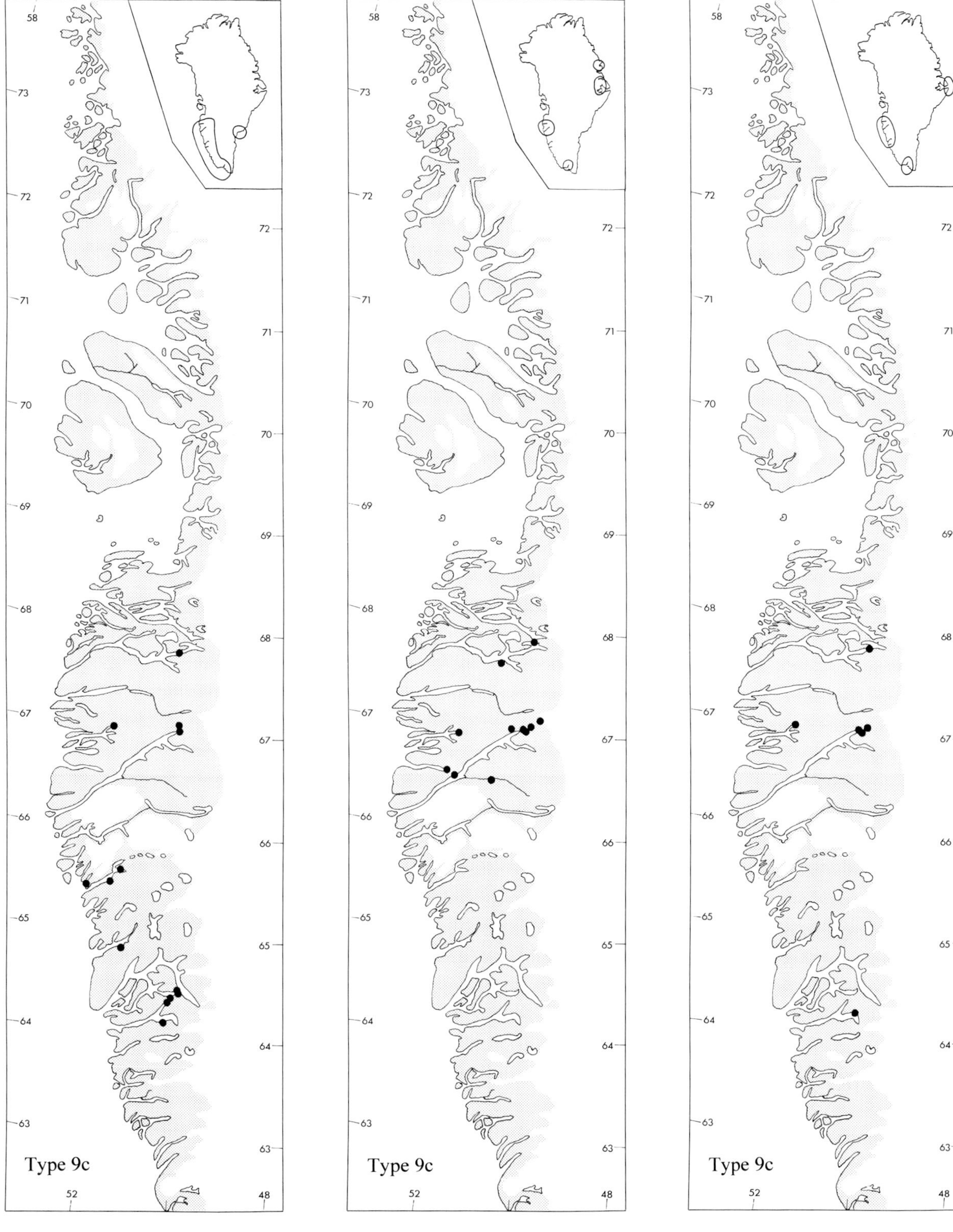

277. Galium brandegei

278. Gentiana detonsa

279. Juncus ranarius

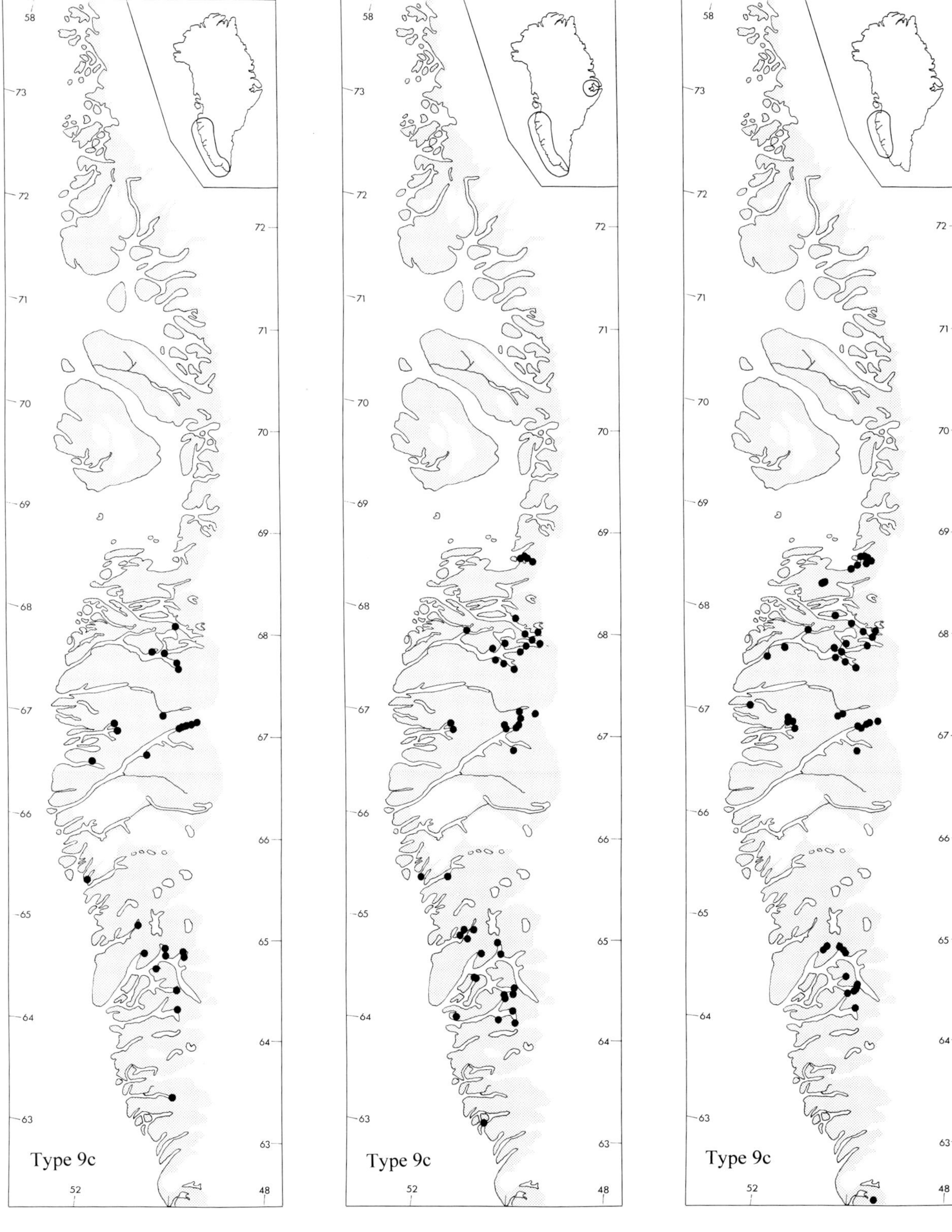

280. Lomatogonium rotatum

281. Menyanthes trifoliata

282. Pedicularis labradorica

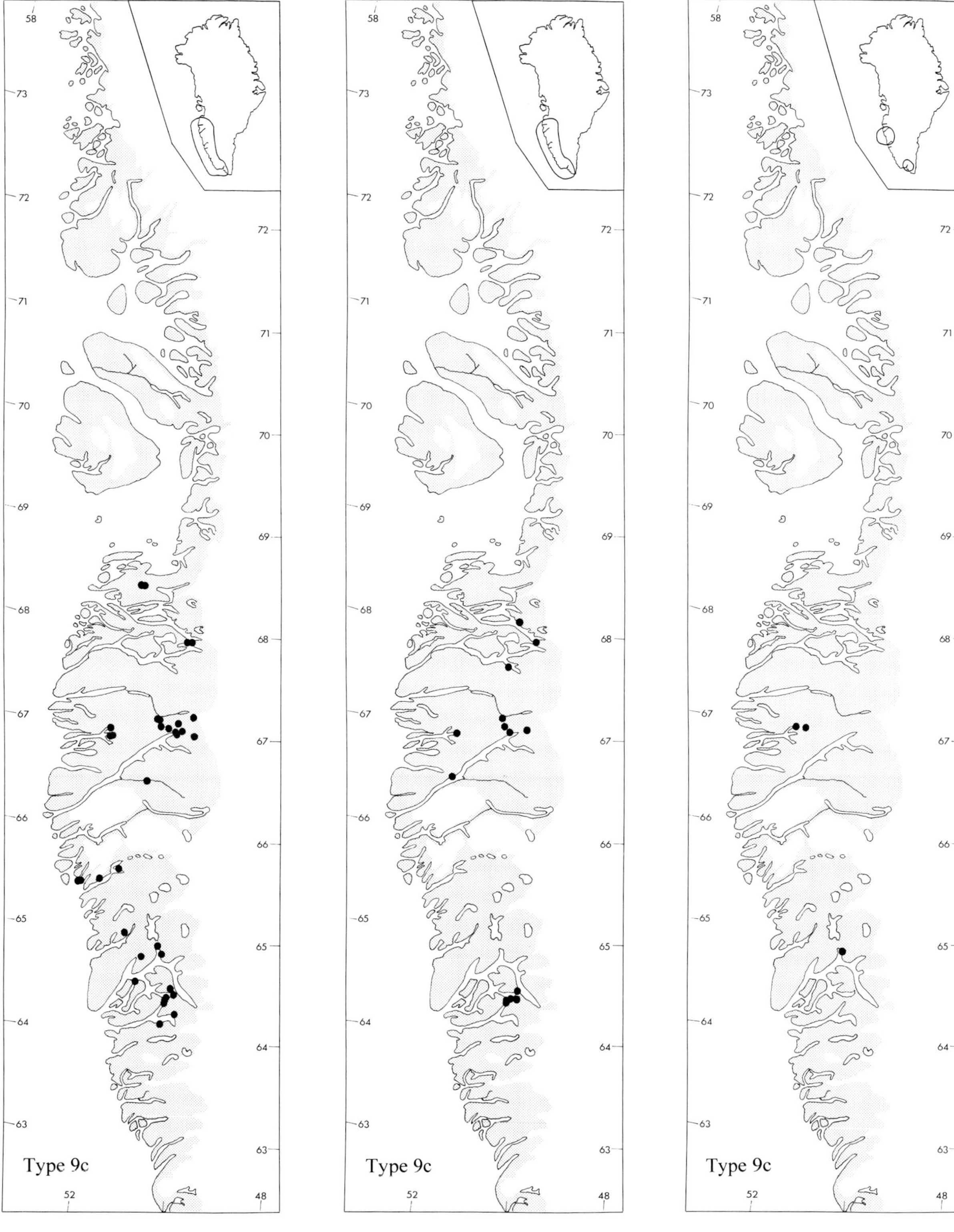

283. Potamogeton alpinus ssp. tenuifolius

284. Potamogeton gramineus

285. Primula egaliksensis

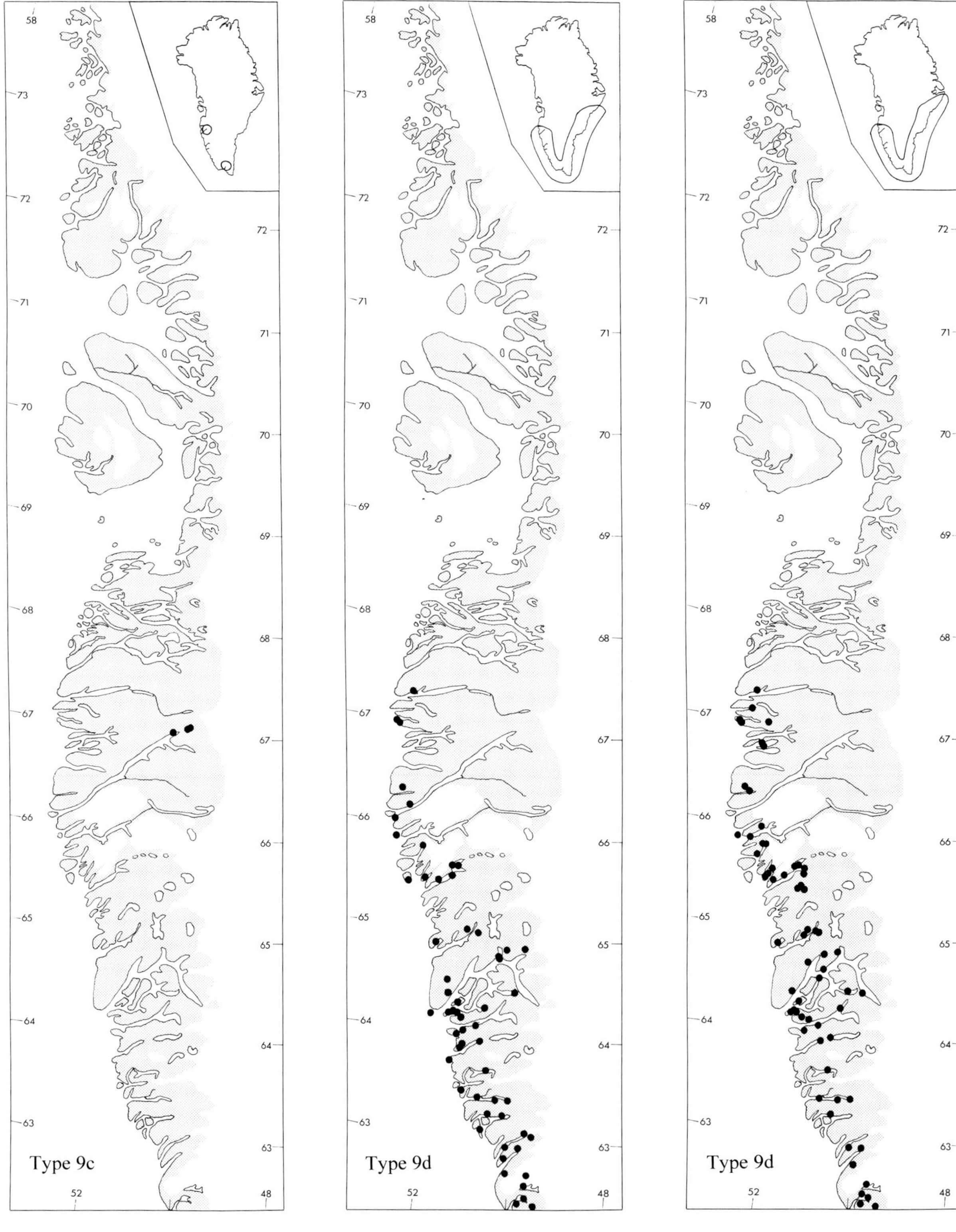

286. Rorippa islandica

287. Alchemilla alpina

288. Alchemilla filicaulis

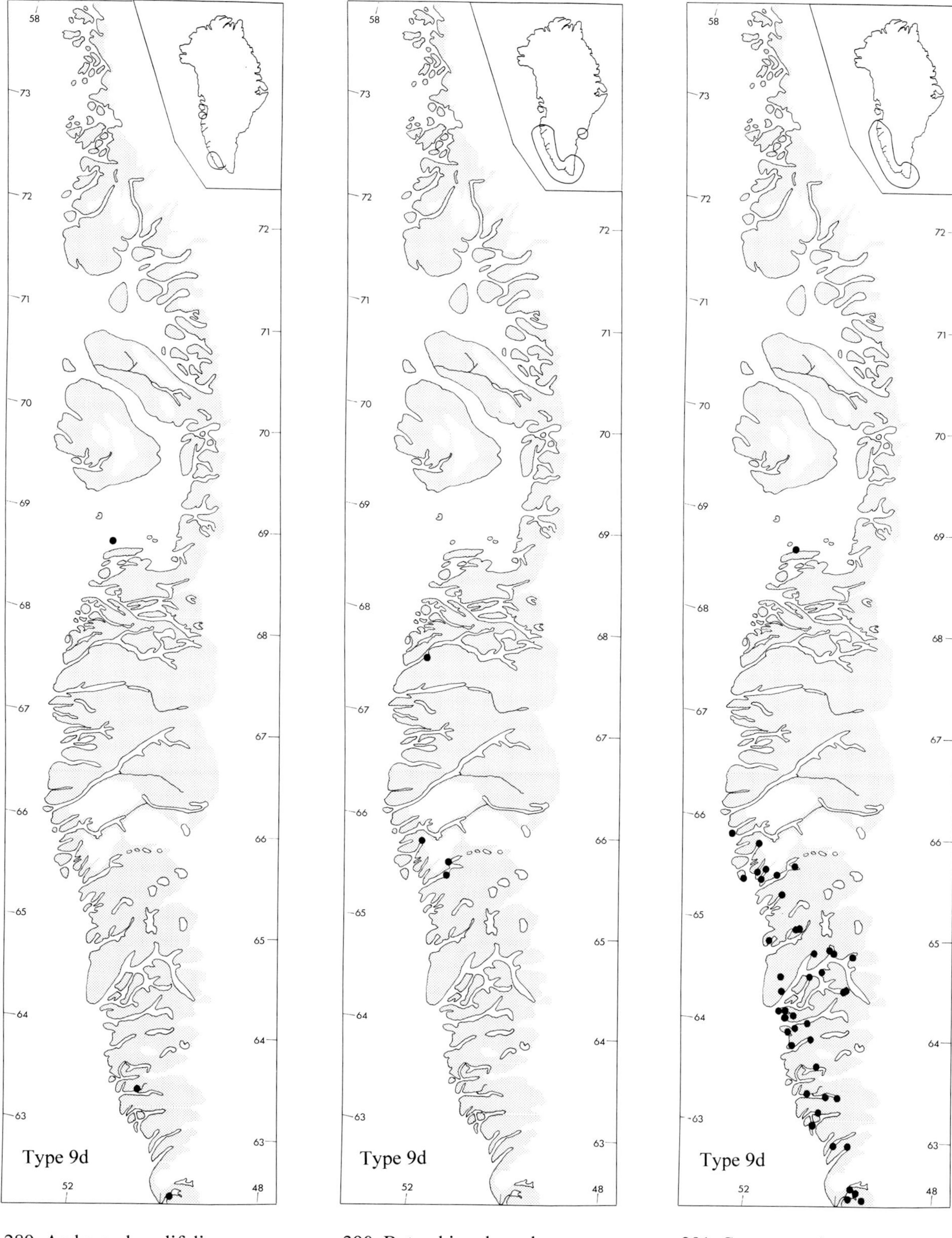

289. Andromeda polifolia

290. Botrychium boreale

291. Cornus suecica

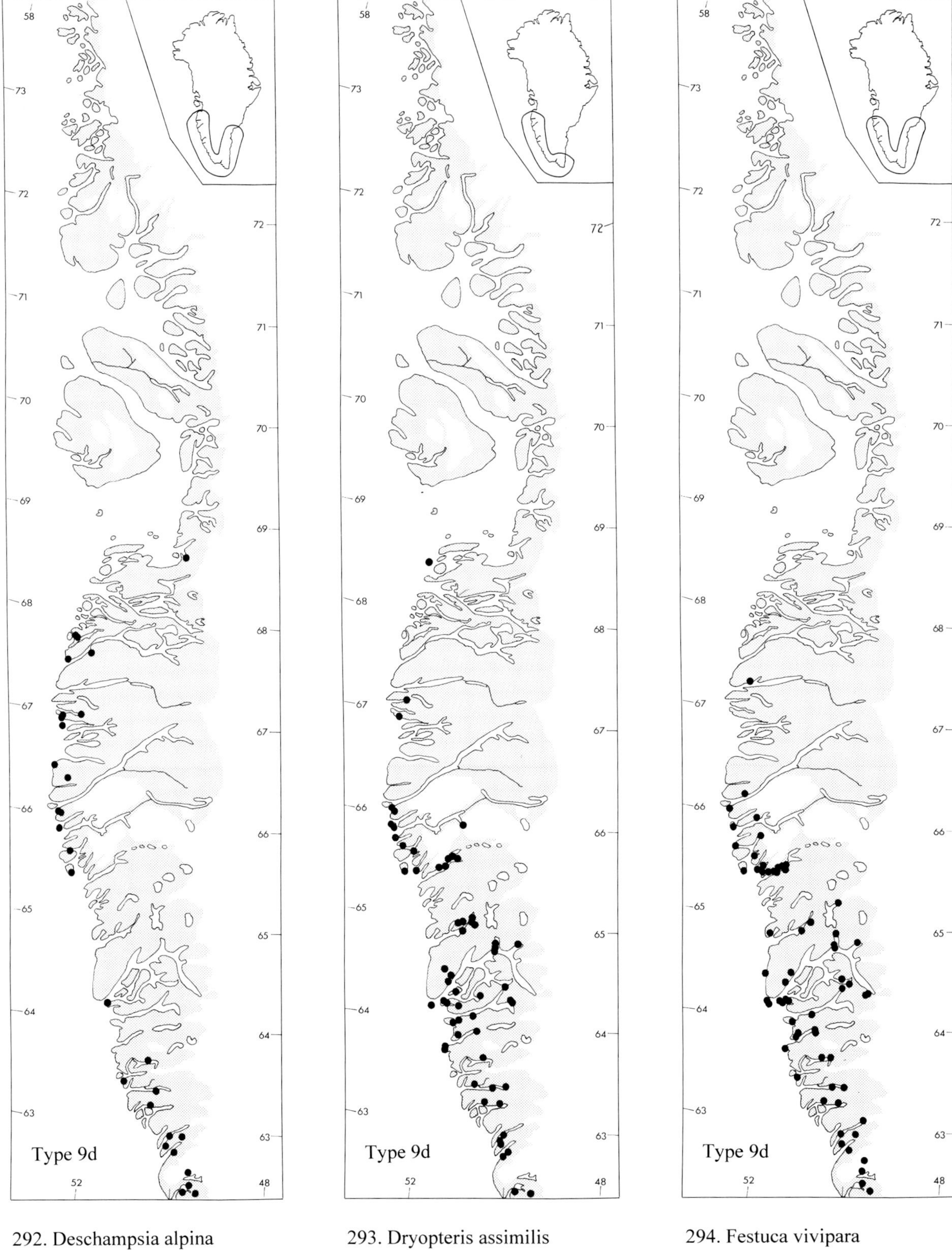

292. Deschampsia alpina

293. Dryopteris assimilis

294. Festuca vivipara var. hirsuta

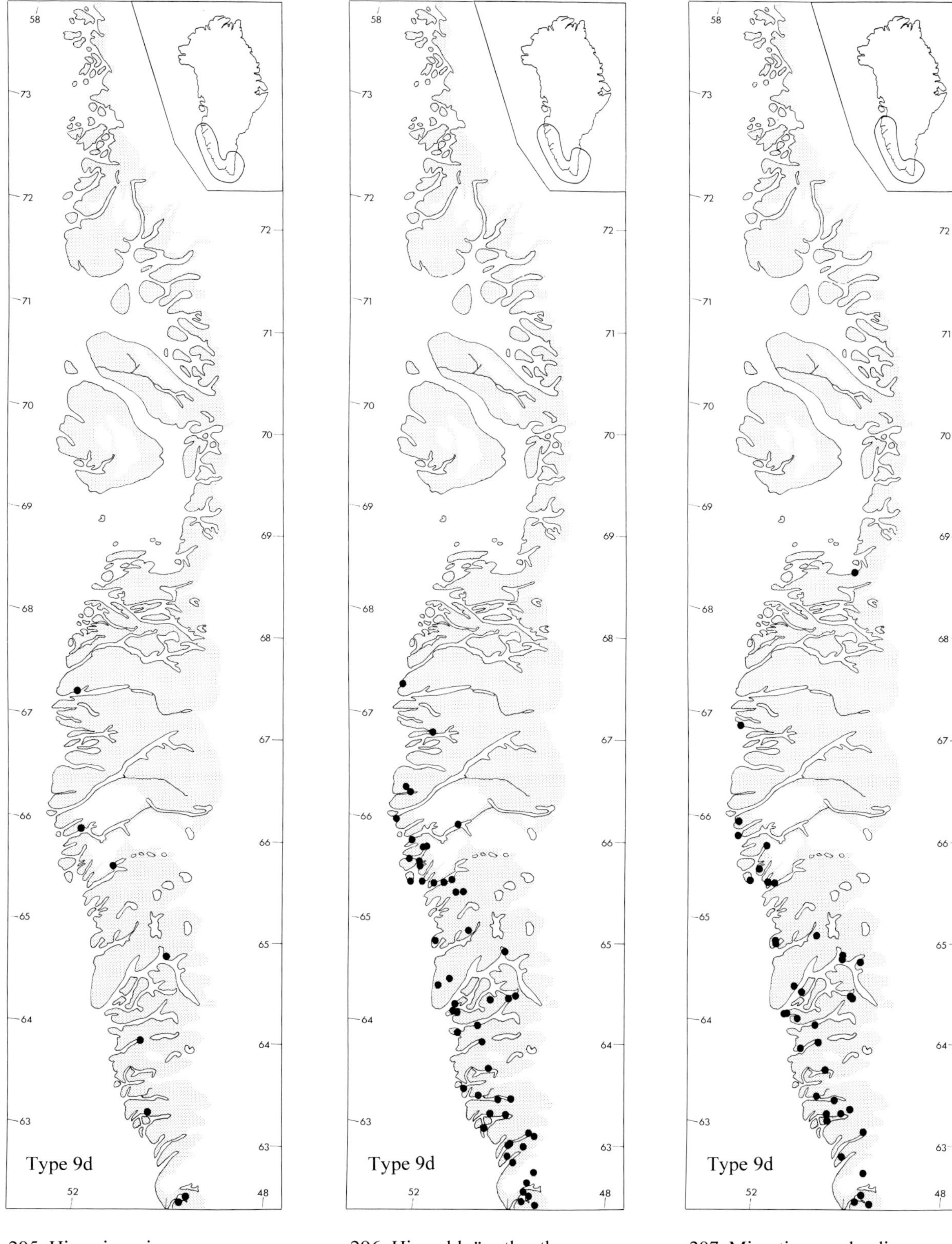

295. Hieracium rigorosum

296. Hierochloë orthantha

297. Minartia groenlandica

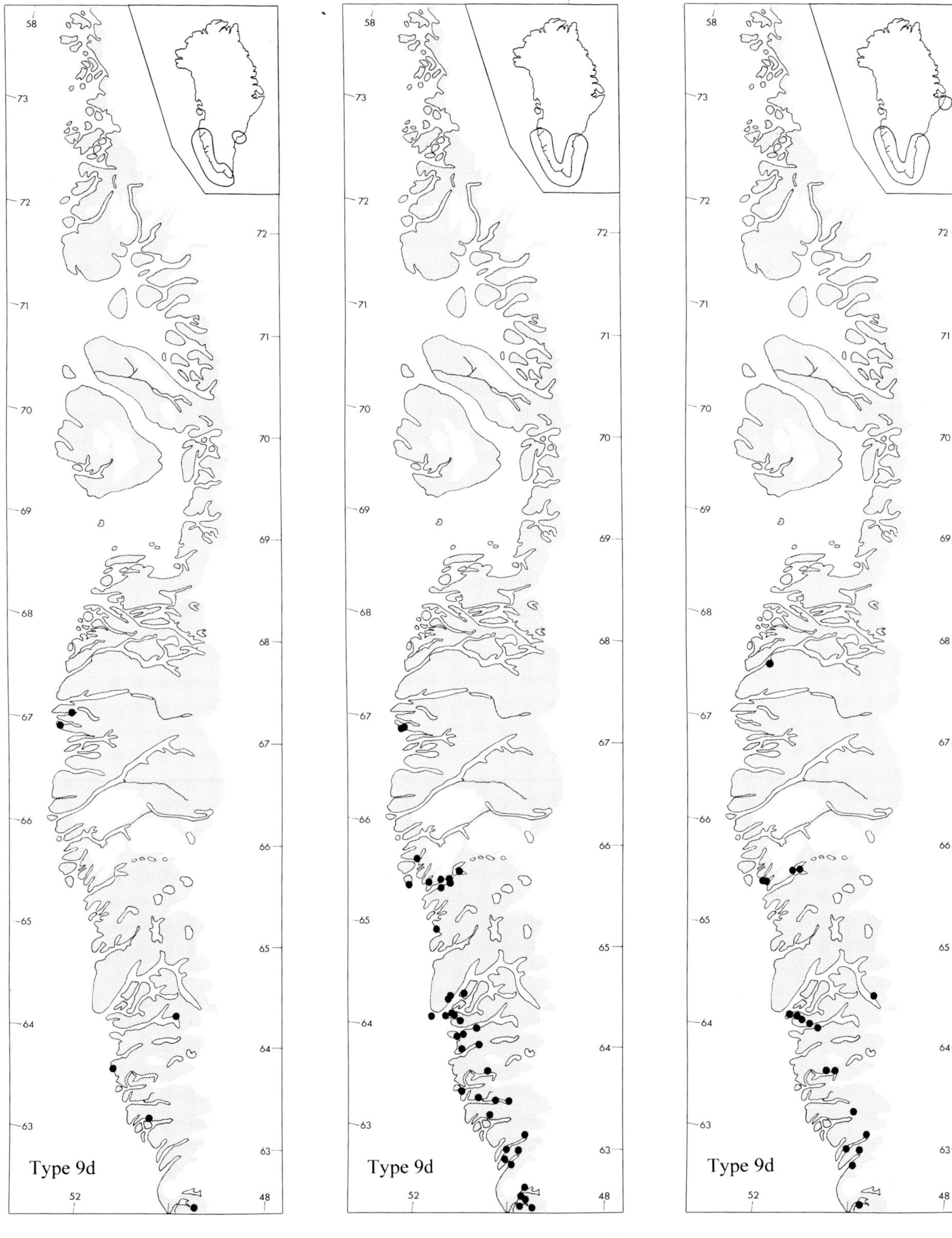

298. Ranunculus acris

299. Saxifraga stellaris

300. Selaginella selaginoides

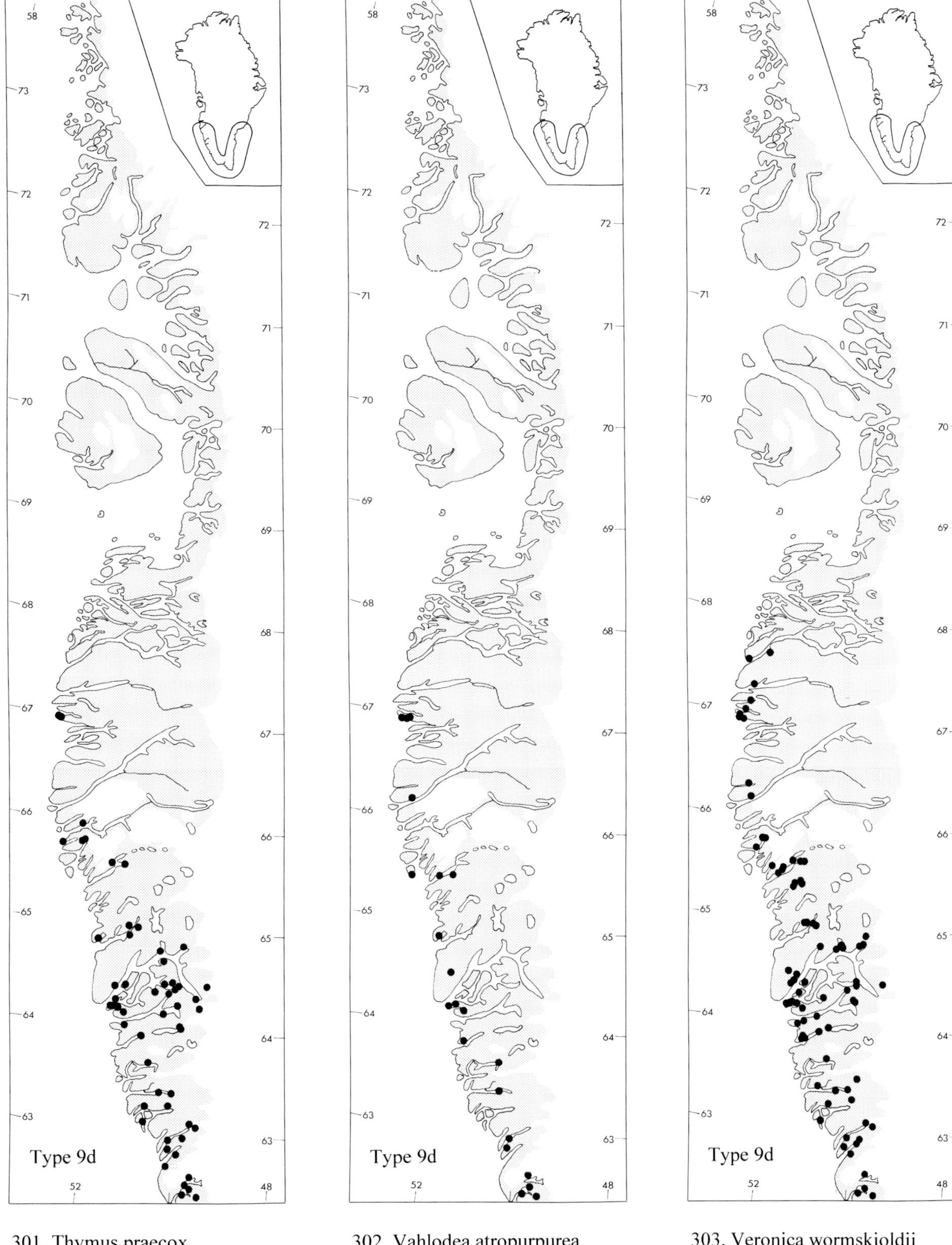

301. Thymus praecox ssp. arcticus

302. Vahlodea atropurpurea

303. Veronica wormskjoldii

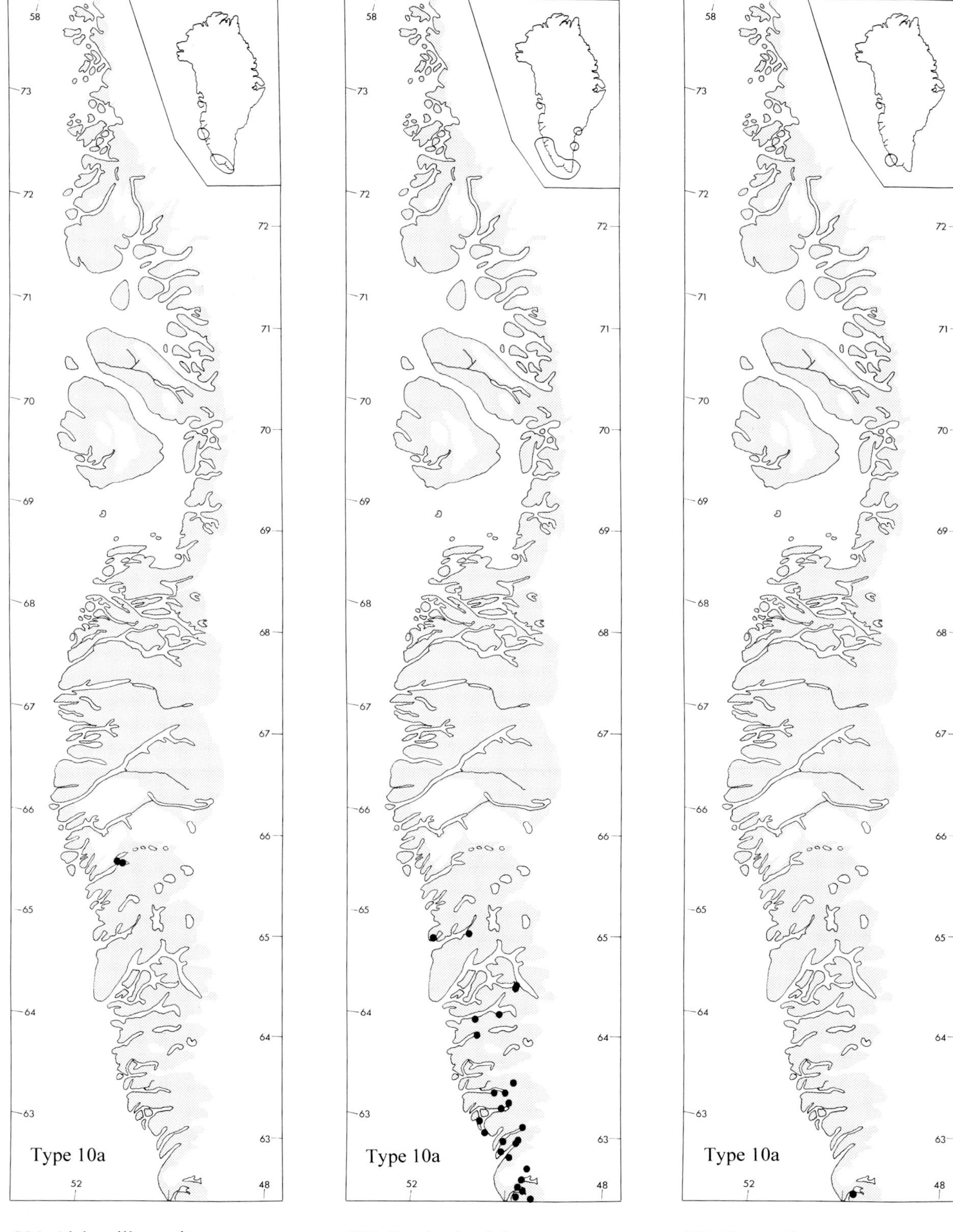

304. Alchemilla vestita

305. Betula glandulosa

306. Carex trisperma

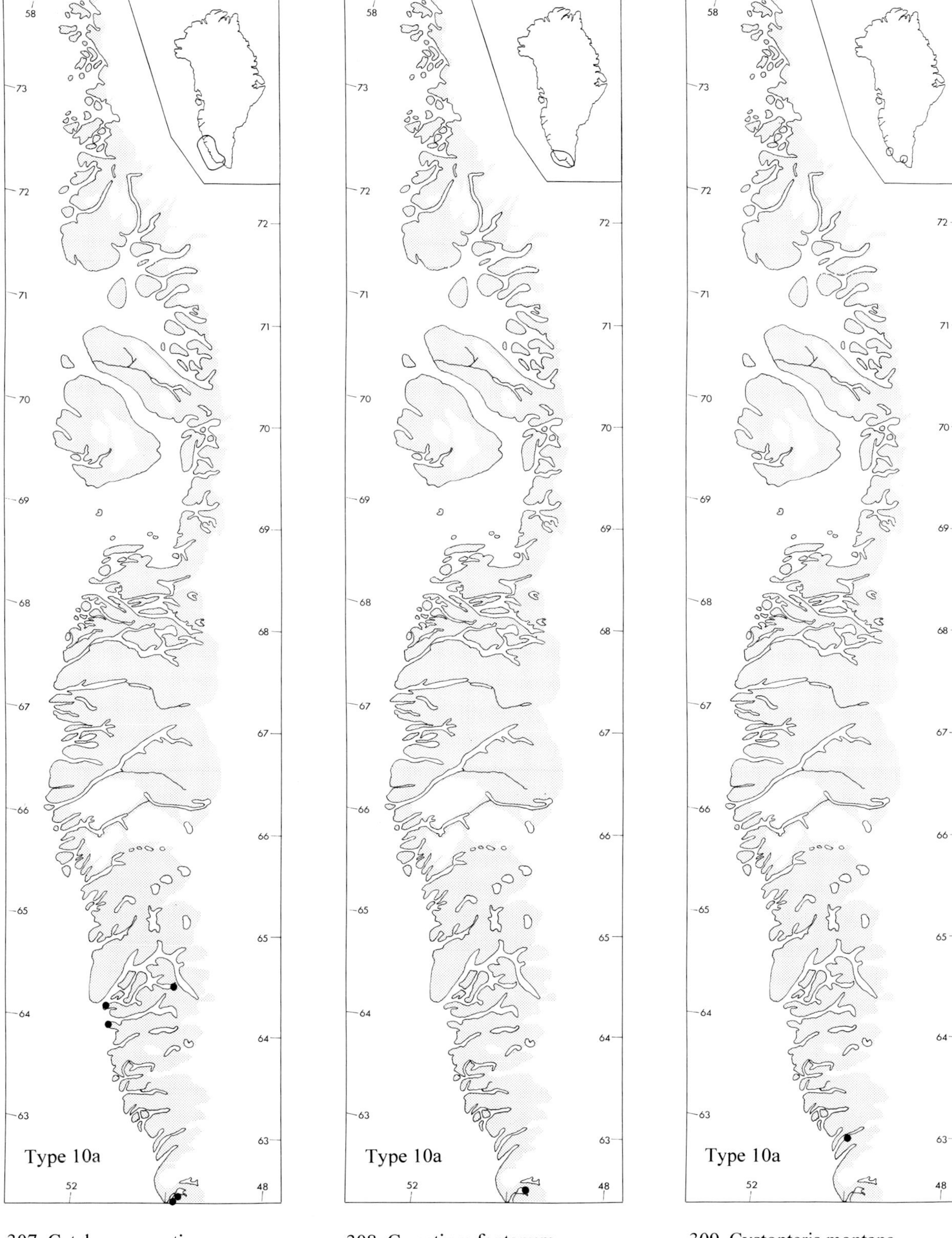

307. Catabrosa aquatica

308. Cerastium fontanum ssp. scandicum

309. Cystopteris montana

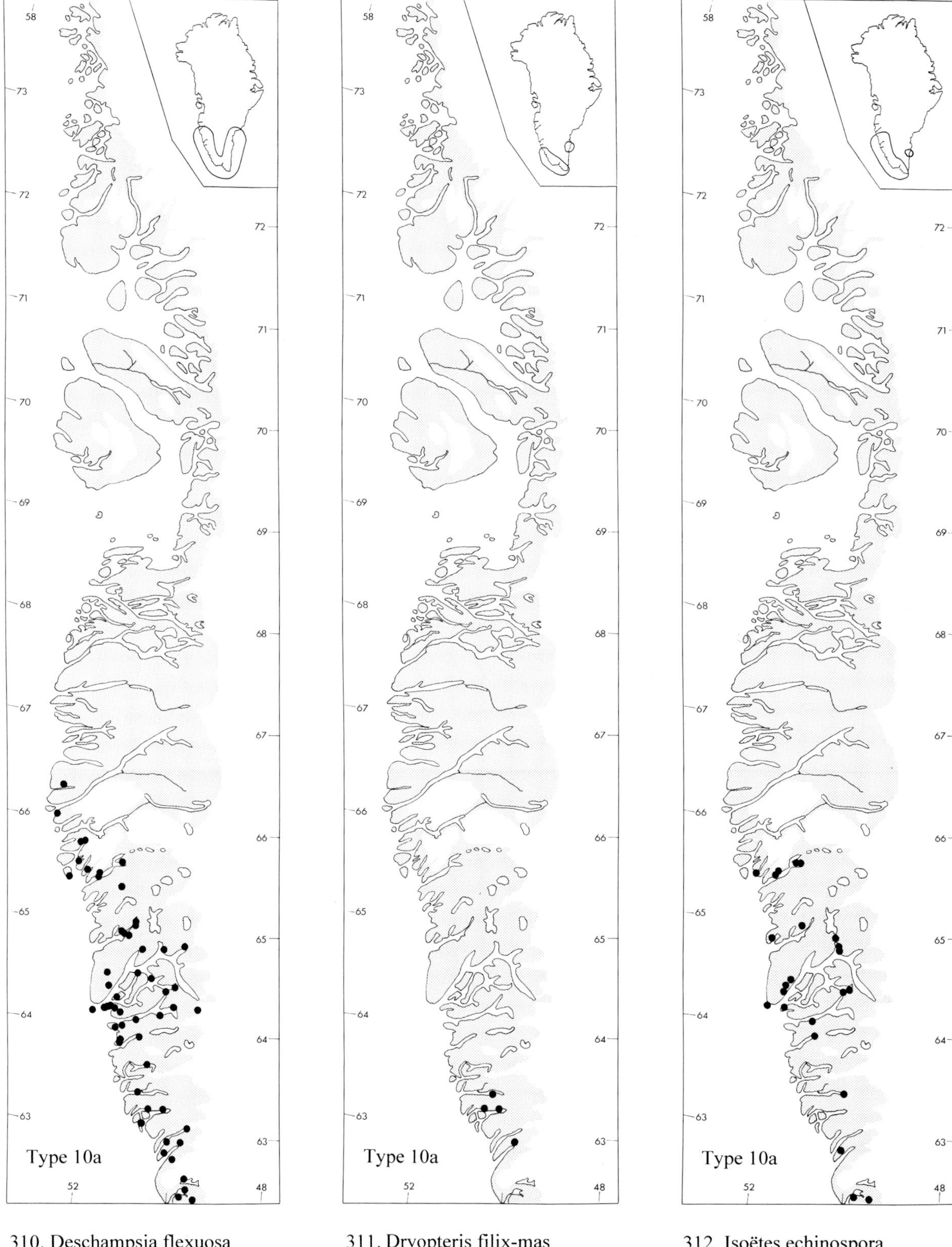

310. Deschampsia flexuosa

311. Dryopteris filix-mas

312. Isoëtes echinospora ssp. muricata

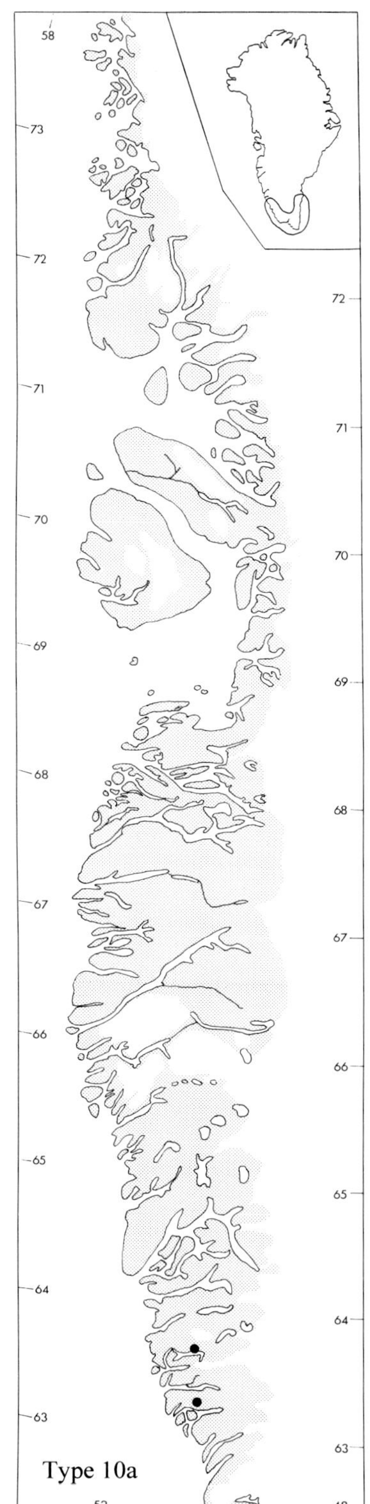

313. Juncus filiformis

314. Lycopodium clavatum
ssp. monostachyon

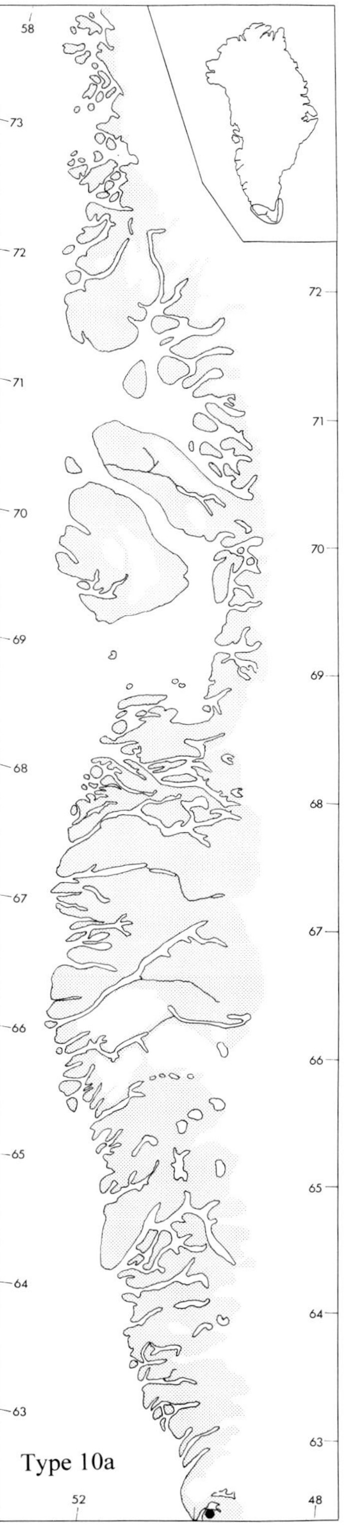

315. Nardus stricta

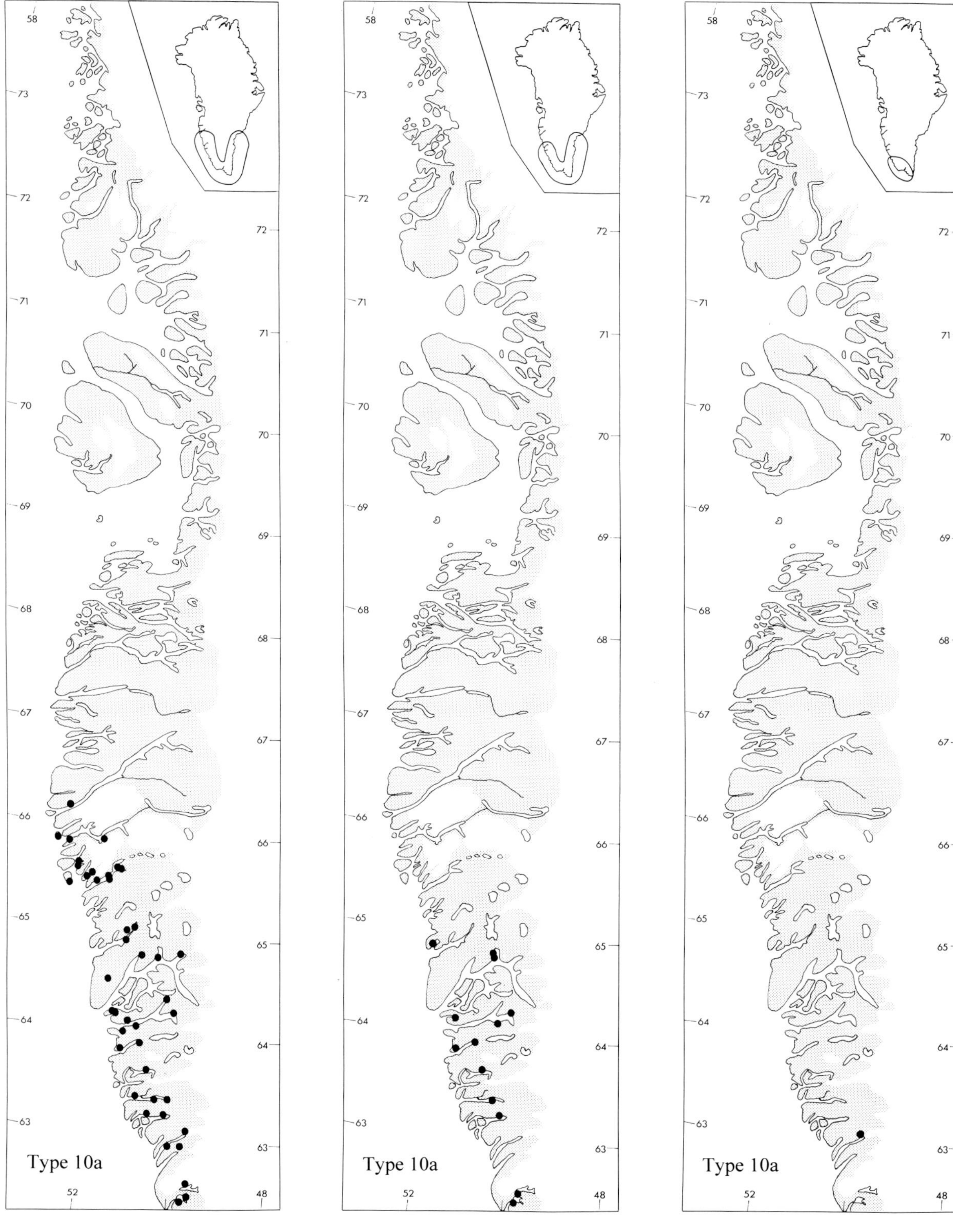

316. Phegopteris connectilis

317. Poa nemoralis

318. Puccinellia maritima

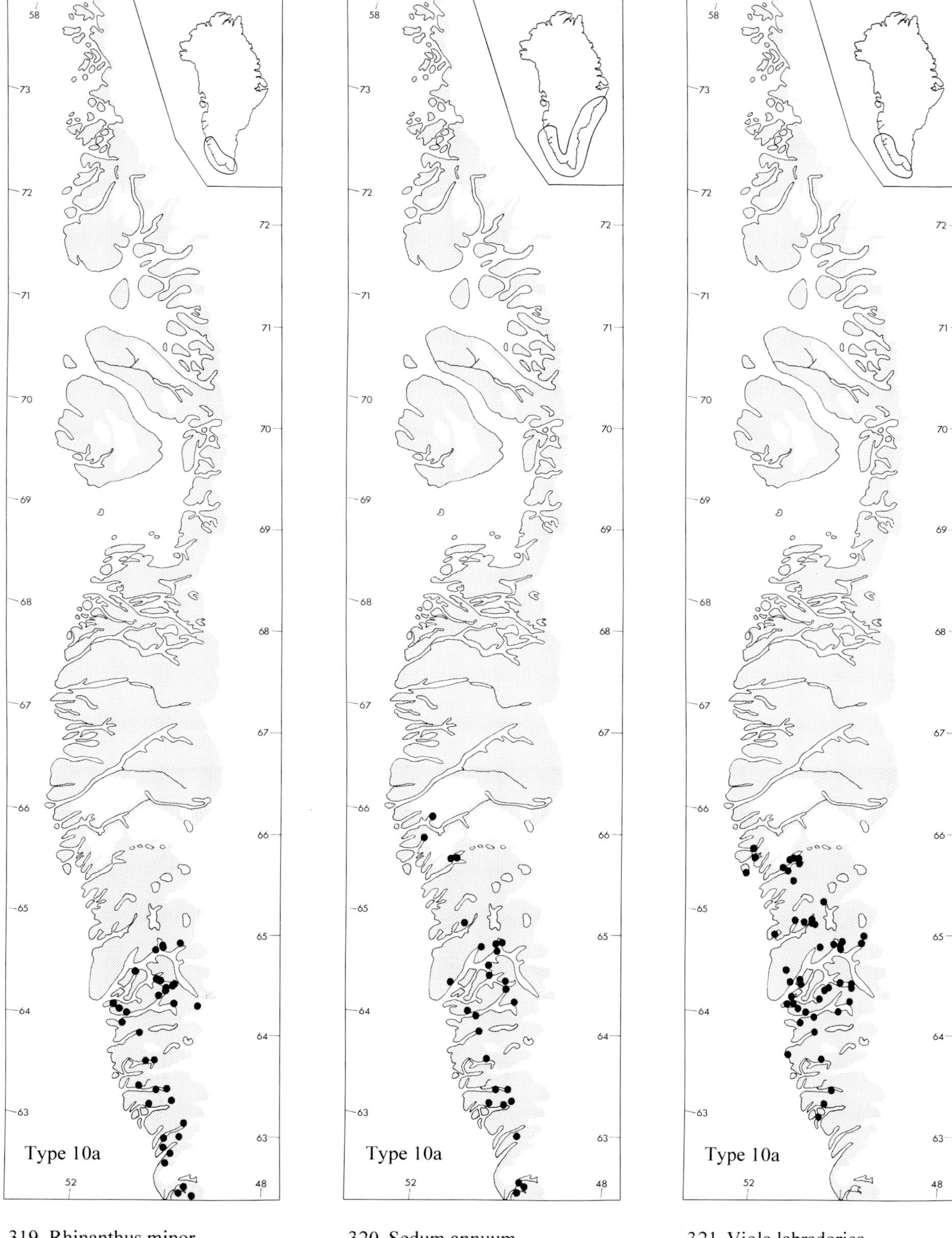

319. Rhinanthus minor

320. Sedum annuum

321. Viola labradorica

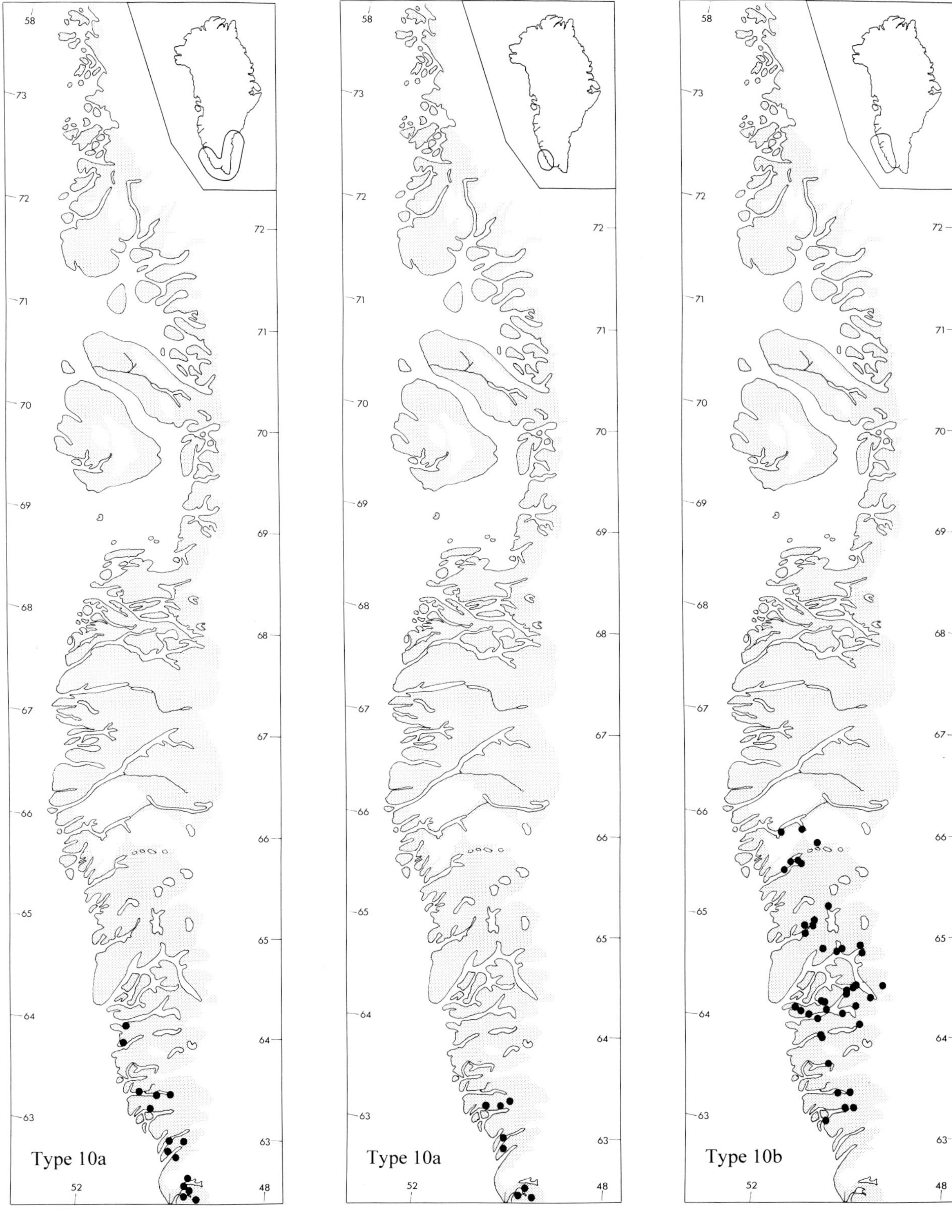

322. Viola palustris

323. Viola selkirkii

324. Alnus crispa

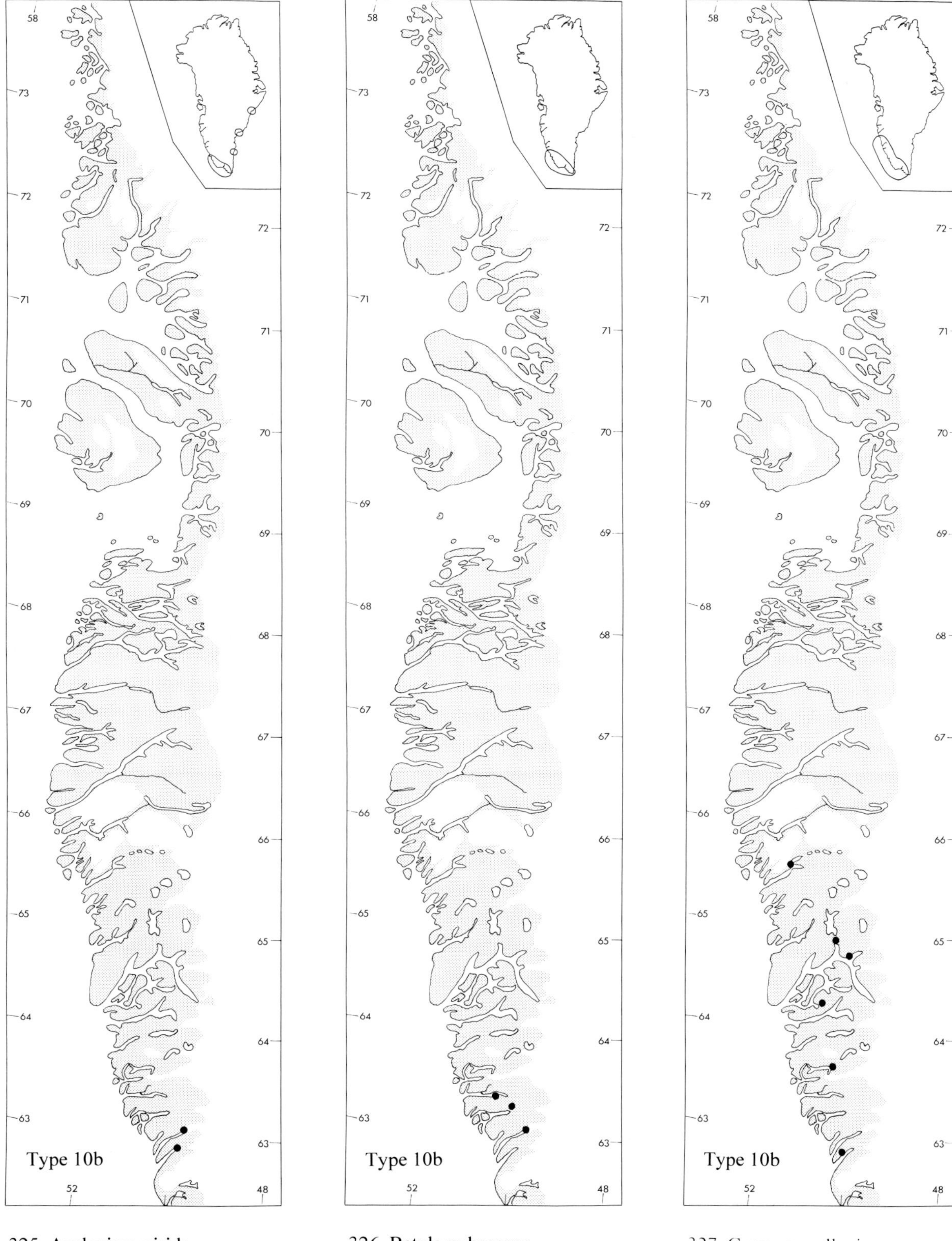

325. Asplenium viride

326. Betula pubescens

327. Carex magellanica
ssp. irrigua

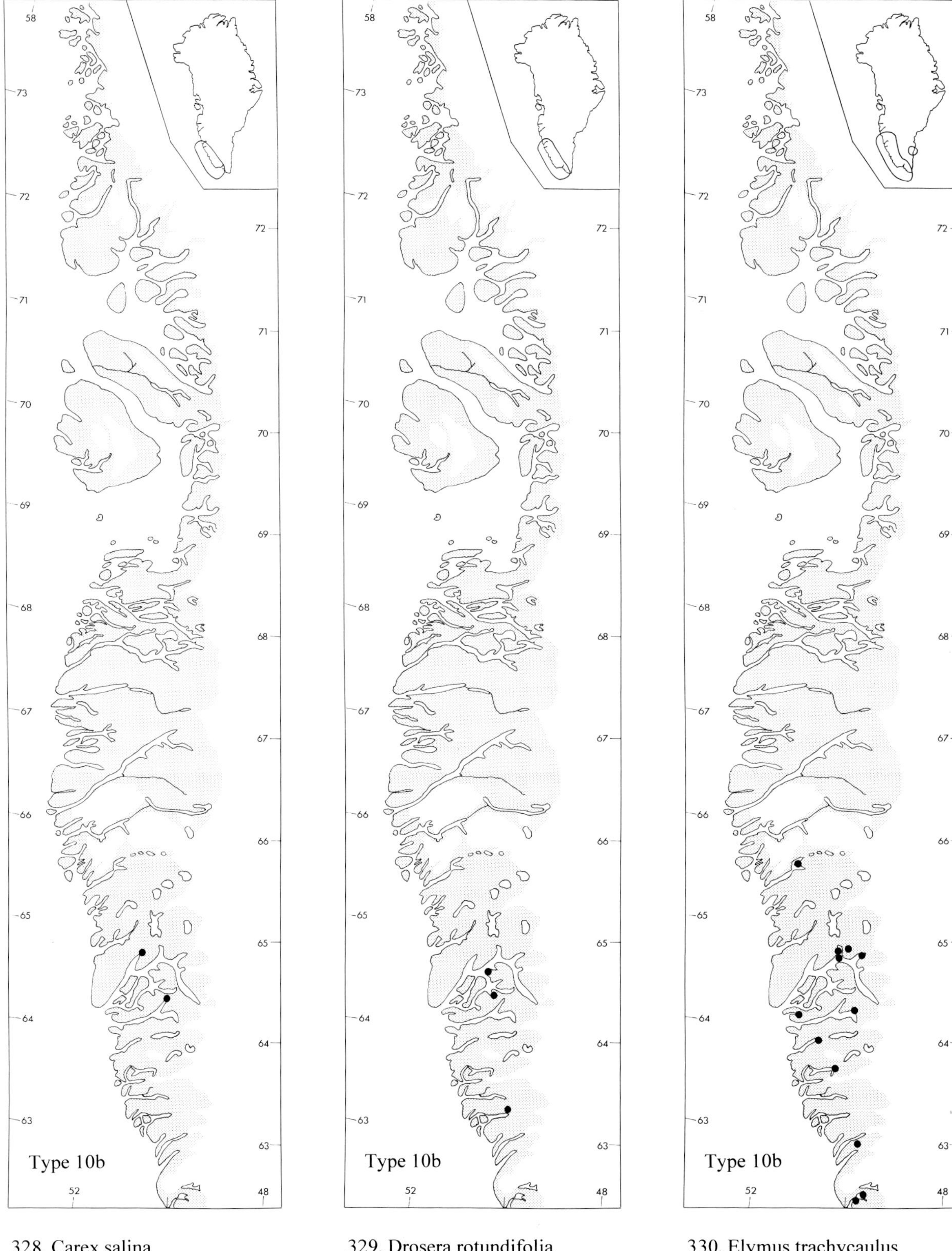

328. Carex salina

329. Drosera rotundifolia

330. Elymus trachycaulus ssp. virescens

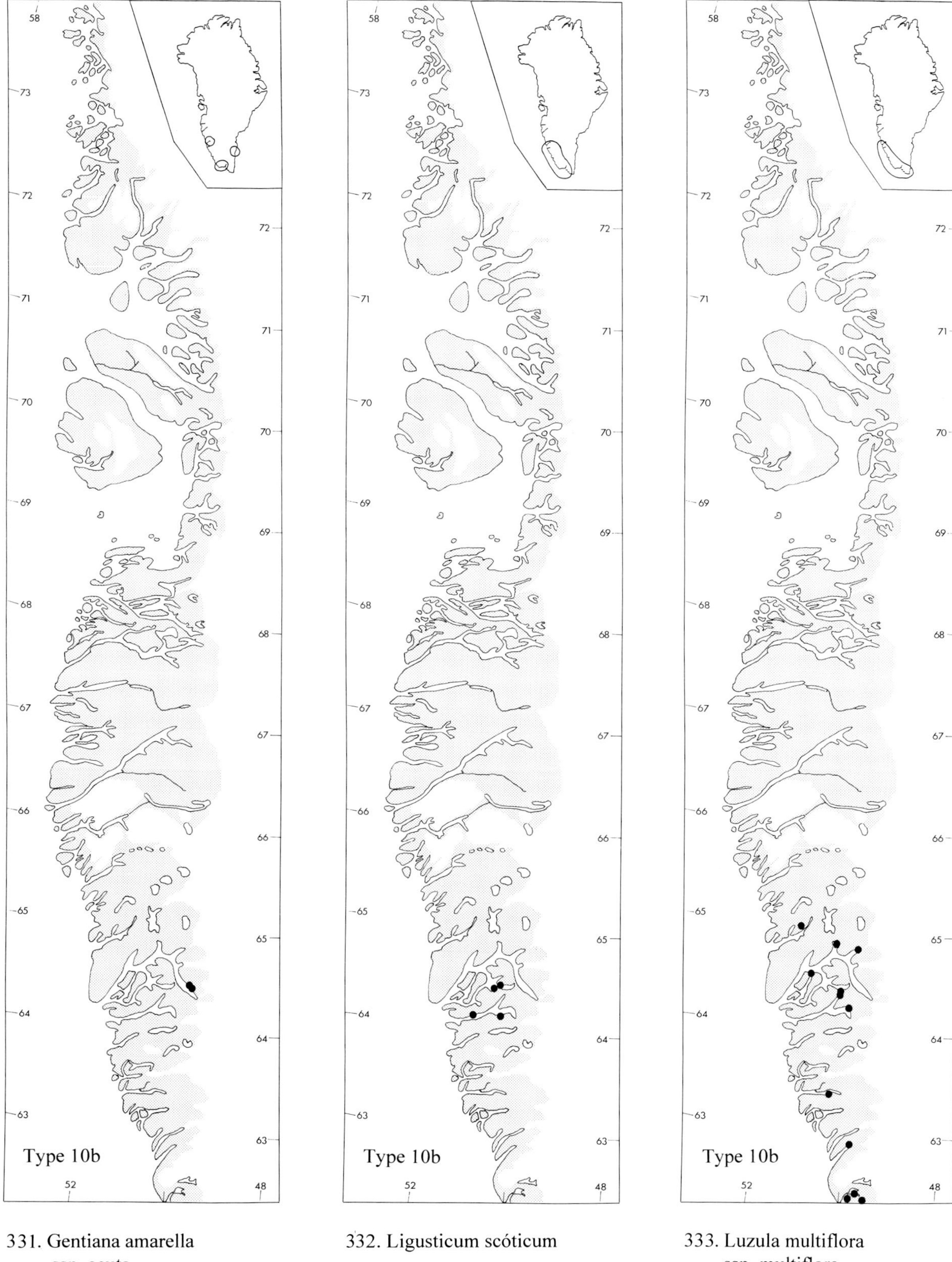

331. Gentiana amarella ssp. acuta

332. Ligusticum scóticum

333. Luzula multiflora ssp. multiflora

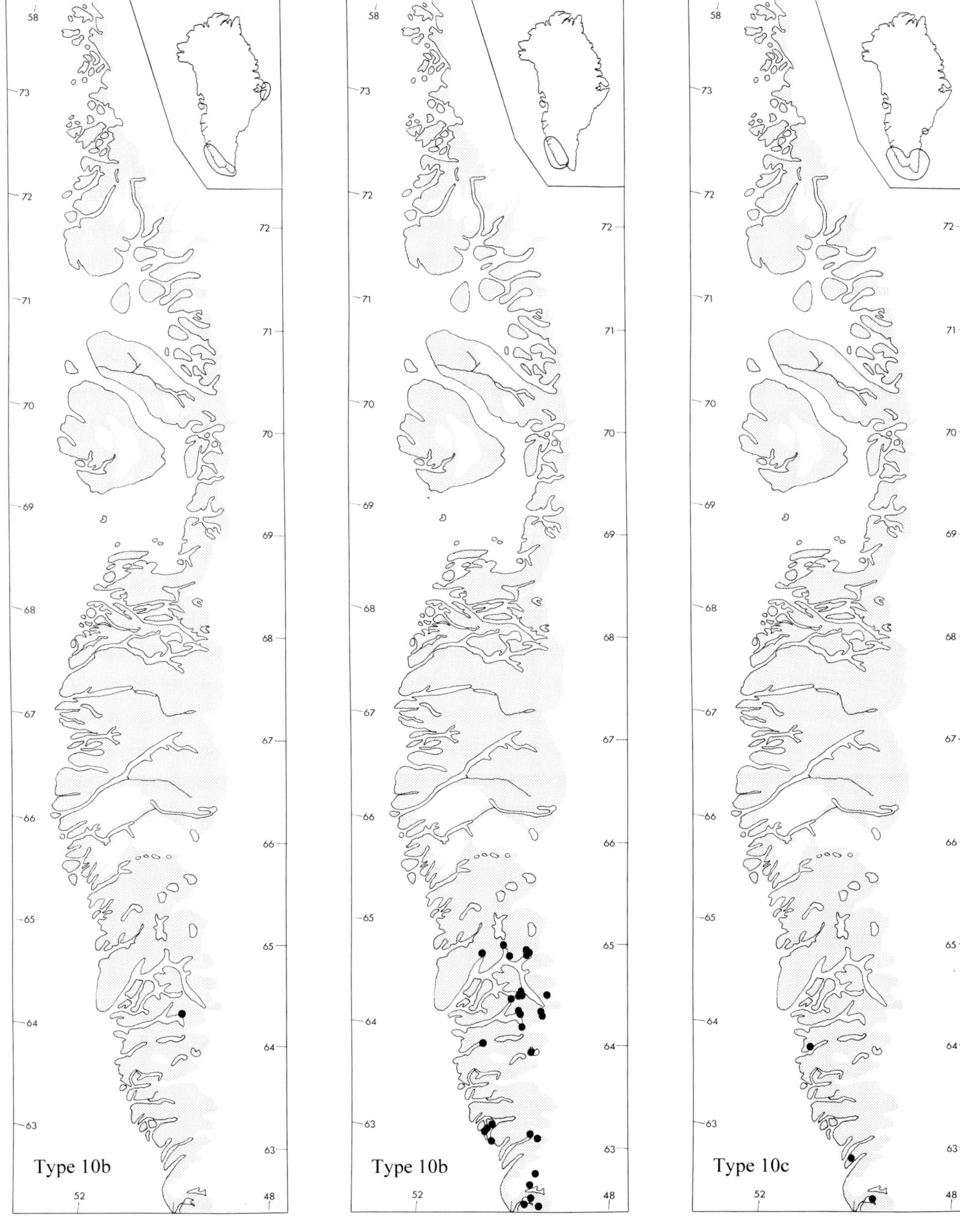

334. Sagina procumbens

335. Salix uva-ursi

336. Athyrium distentifolium

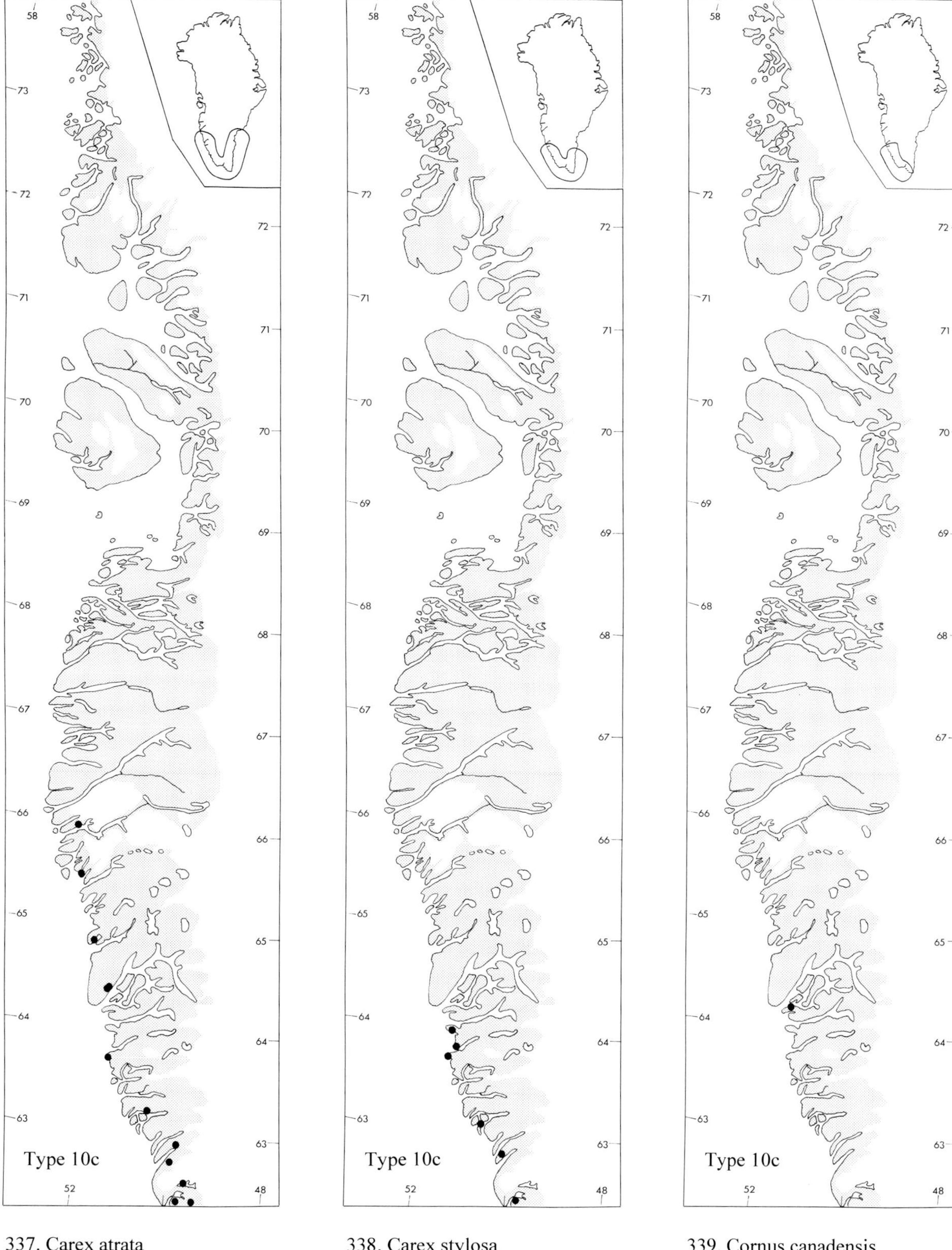

337. Carex atrata

338. Carex stylosa var. nigritella

339. Cornus canadensis

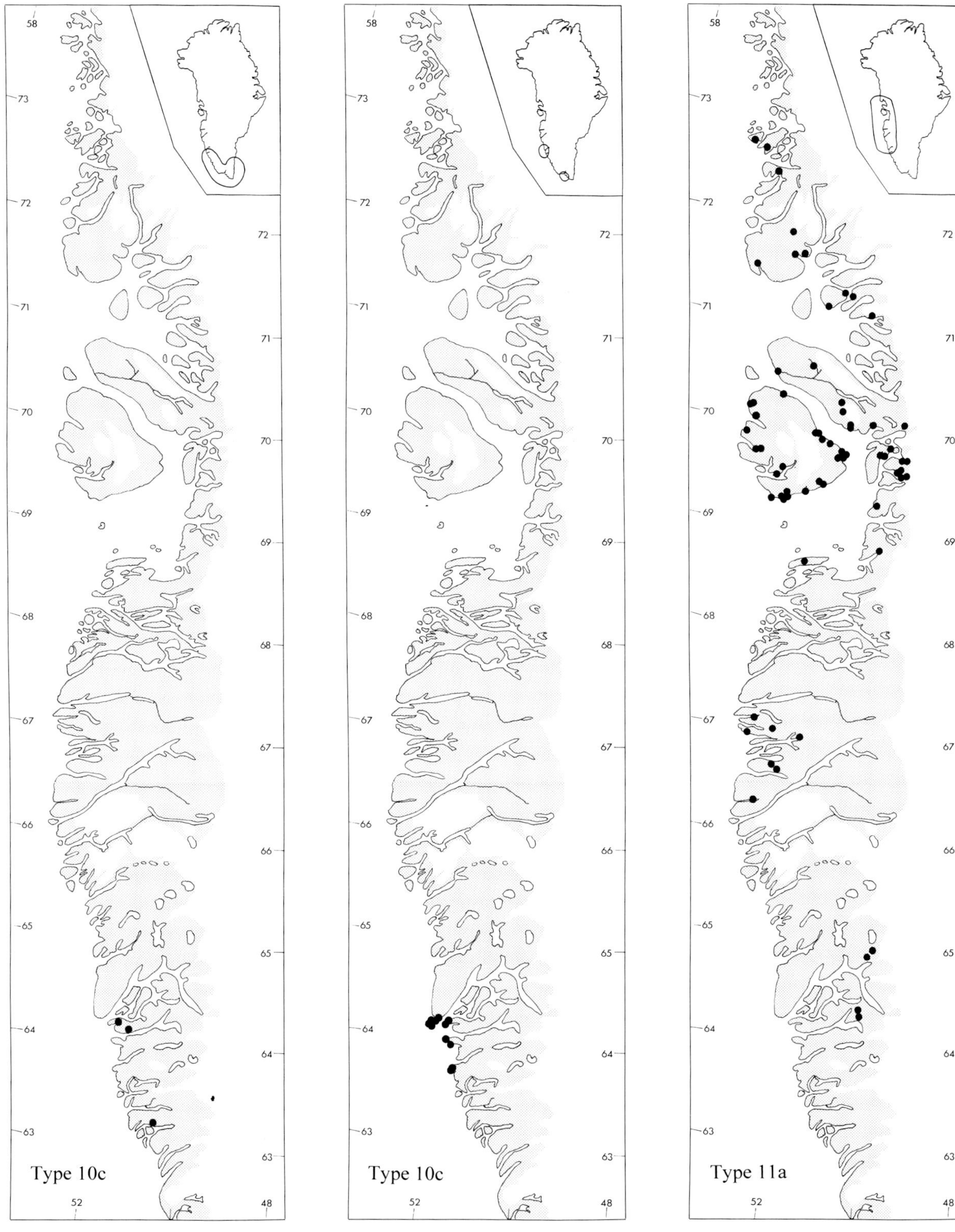

340. Galium triflorum

341. Rubus chamaemorus

342. Antennaria glabrata

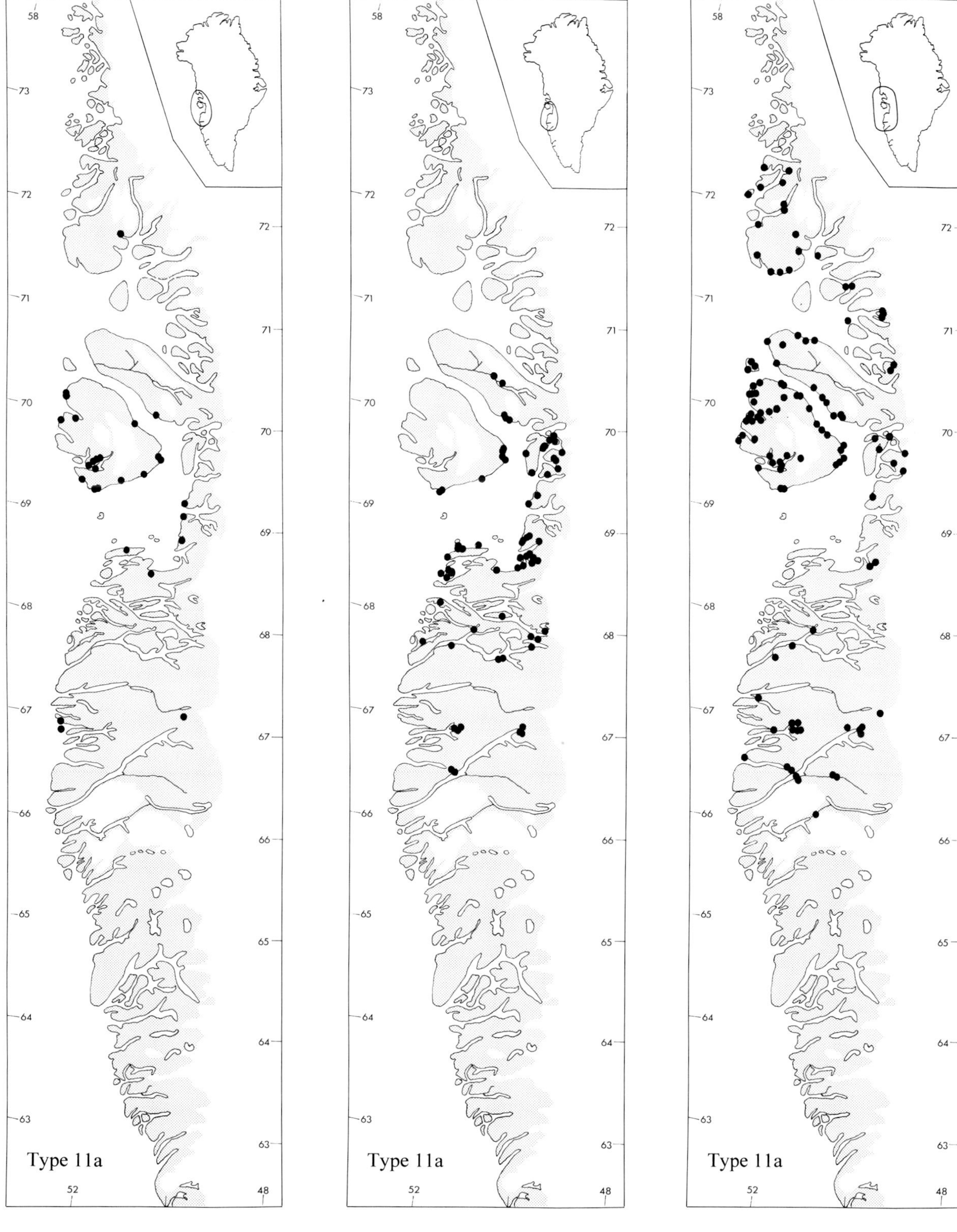

343. Callitriche hermaphroditica

344. Carex holostoma

345. Pedicularis lanata

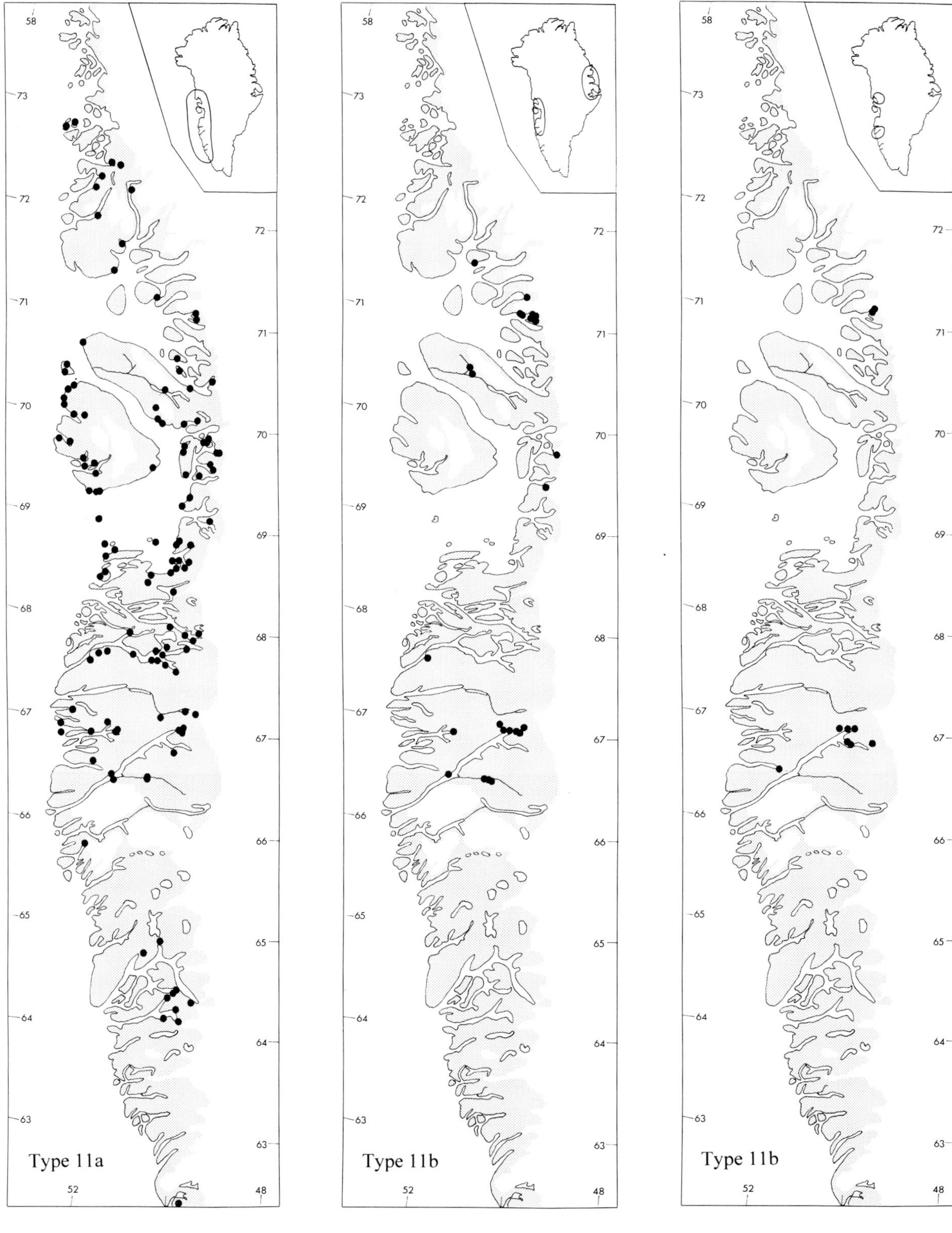

346. Ranunculus lapponicus

347. Braya linearis

348. Braya novae-angliae

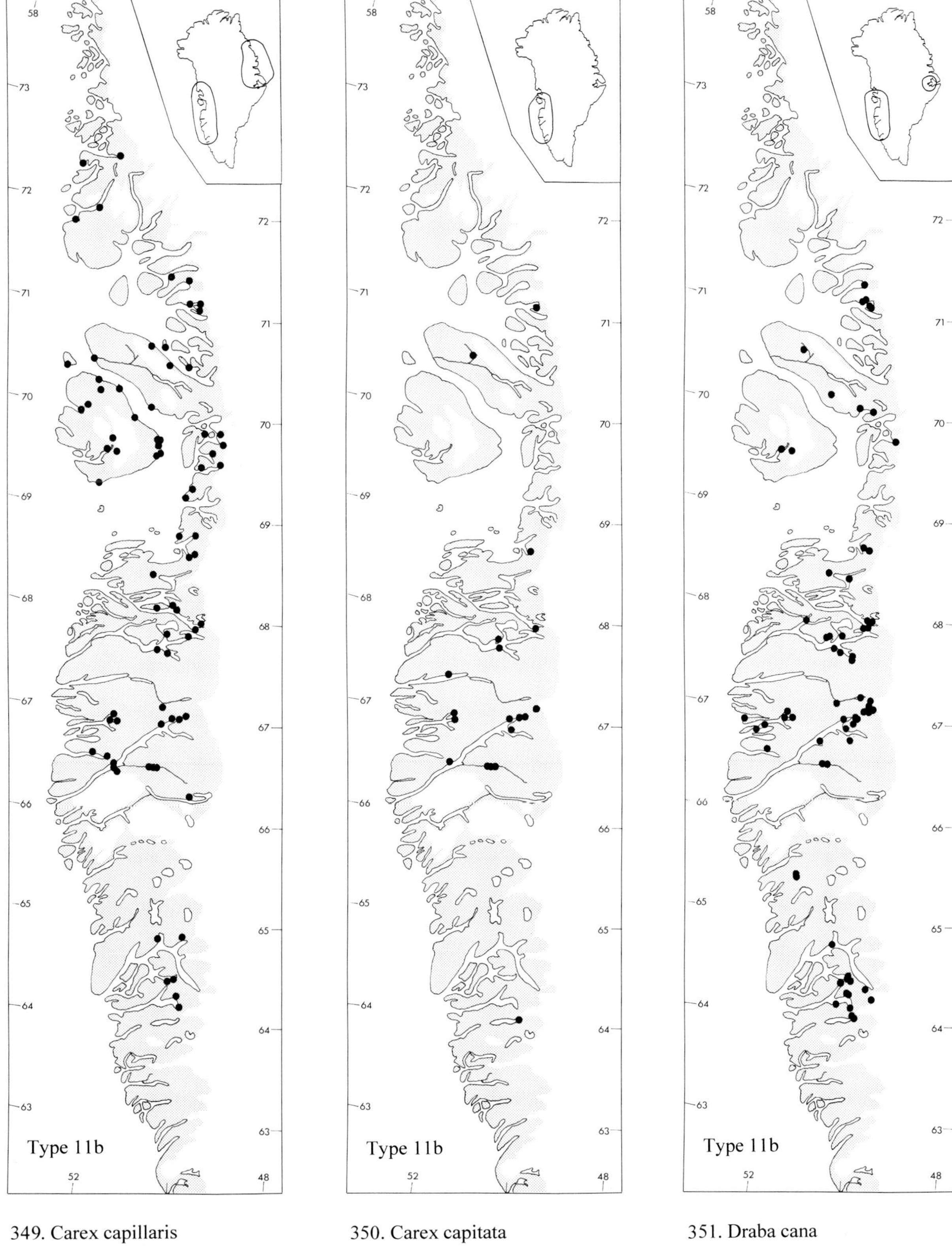

349. Carex capillaris
ssp. robustior

350. Carex capitata
ssp. capitata

351. Draba cana

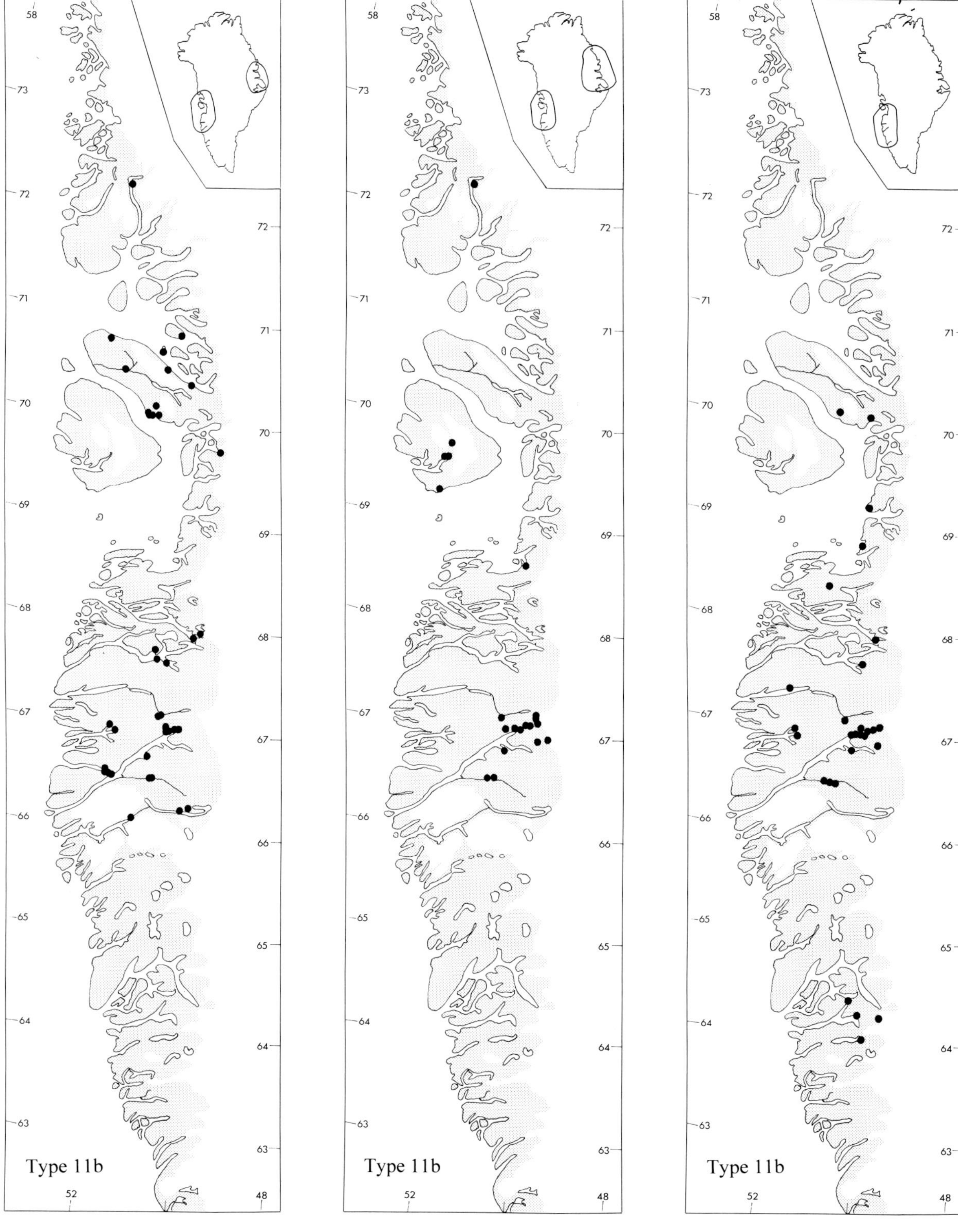

352. Draba cinerea

353. Gentiana tenella

354. Luzula groenlandica

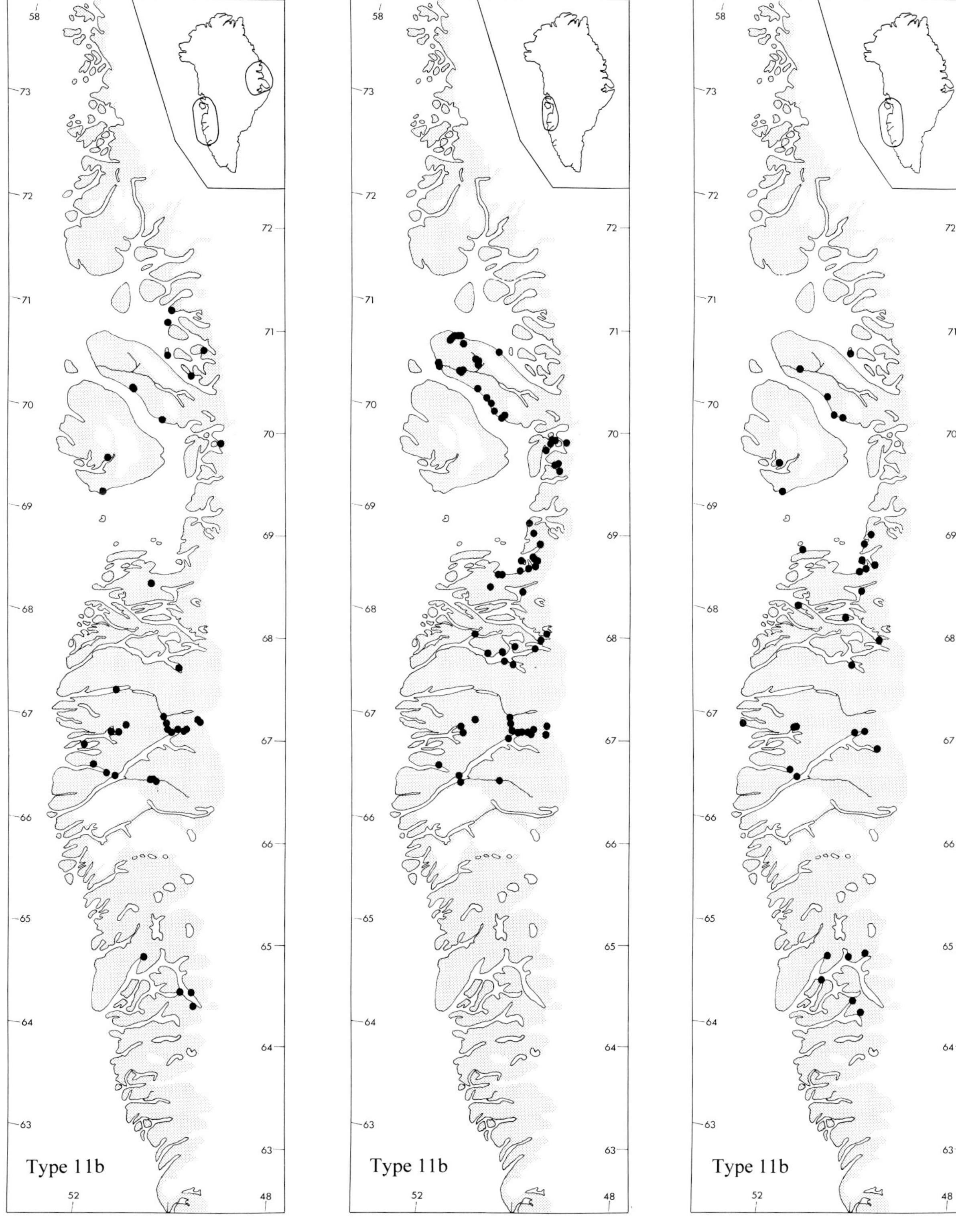

355. Primula stricta

356. Puccinellia deschampsioides

357. Puccinellia groenlandica

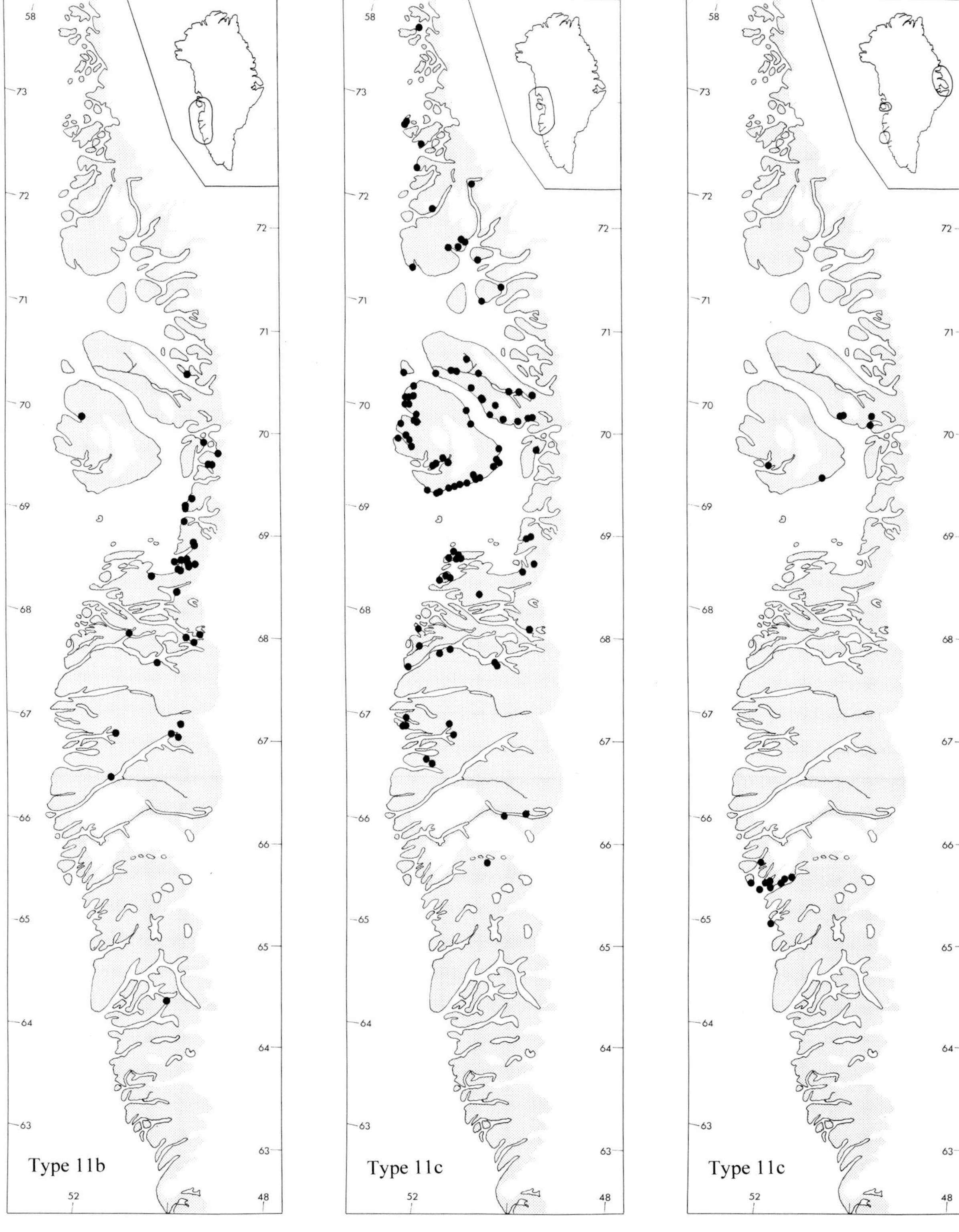

358. Utricularia ochroleuca

359. Antennaria angustata

360. Arctostaphylos alpina

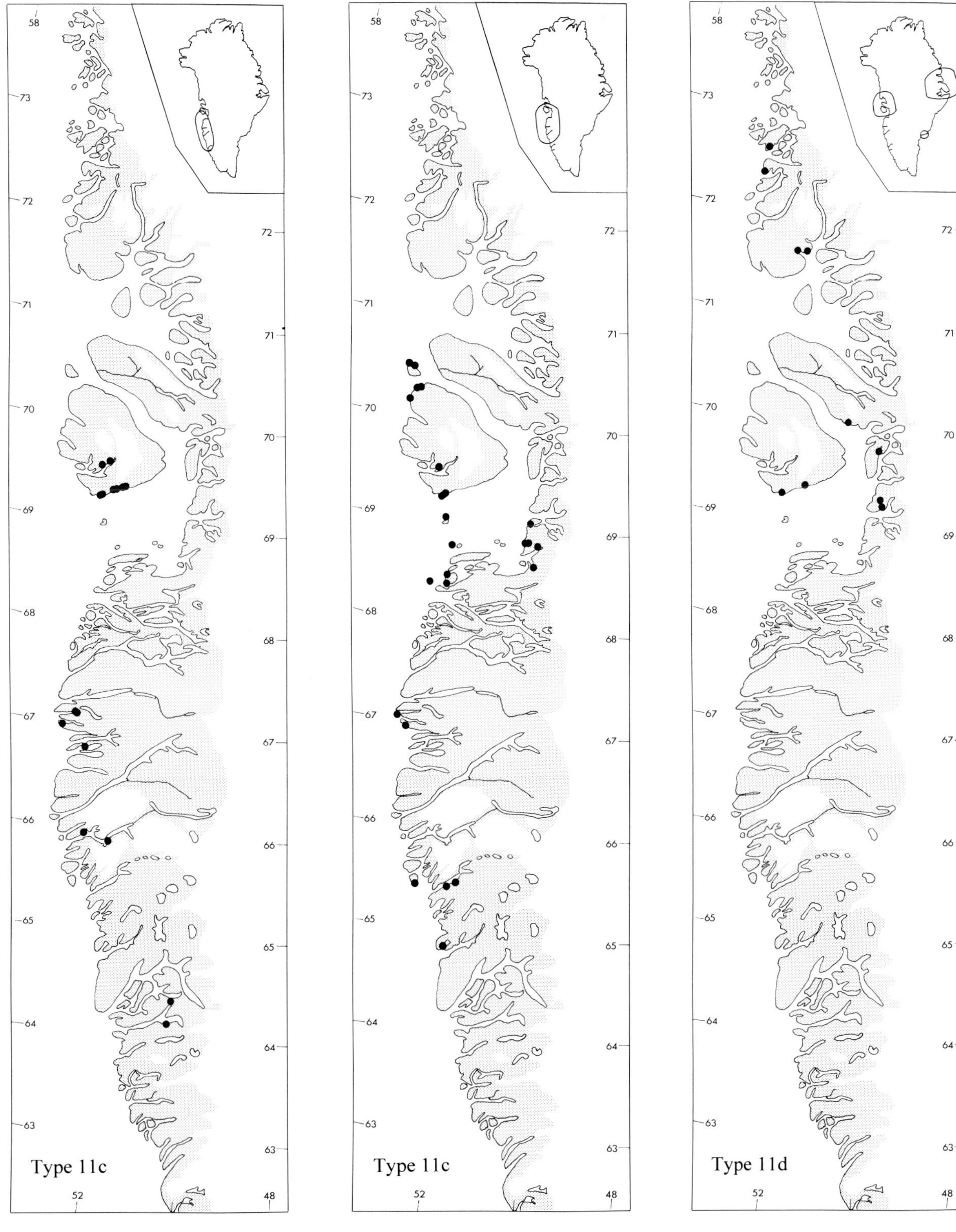

361. Orthilia secunda ssp. obtusata

362. Puccinellia langeana

363. Antennaria porsildii

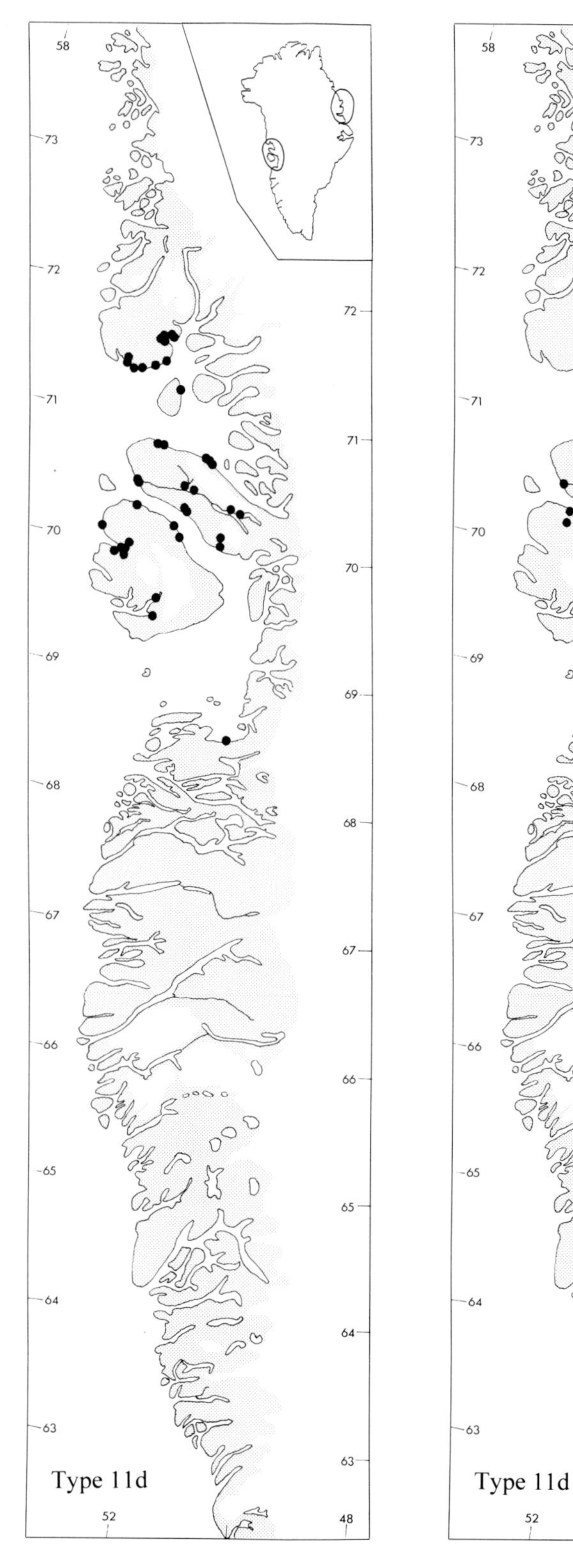

364. Dupontia psilosantha

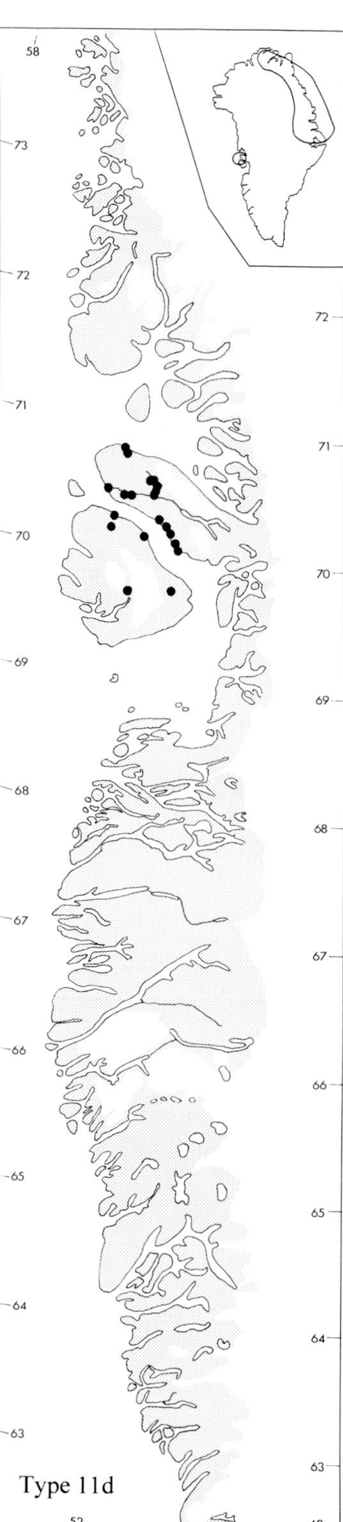

365. Elymus hyperarcticus

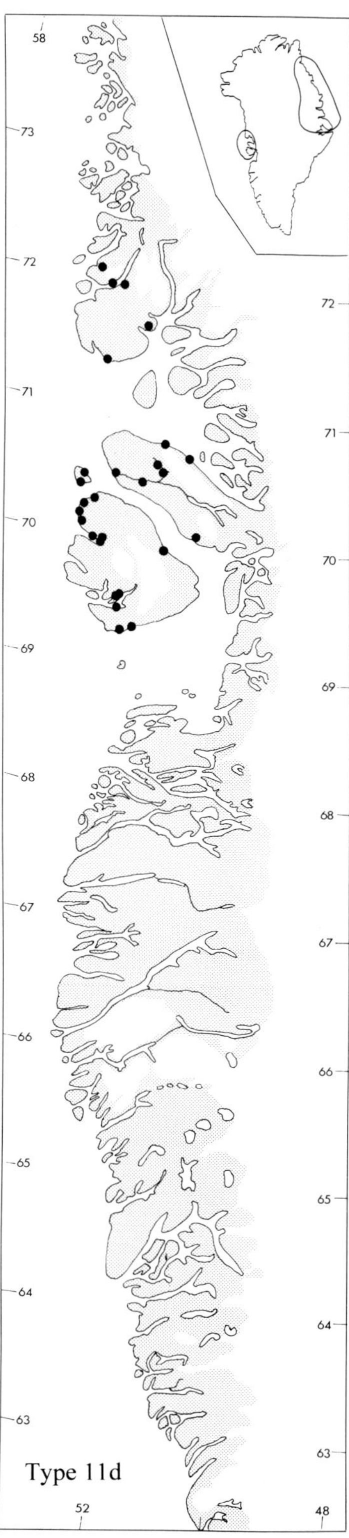

366. Epilobium arcticum

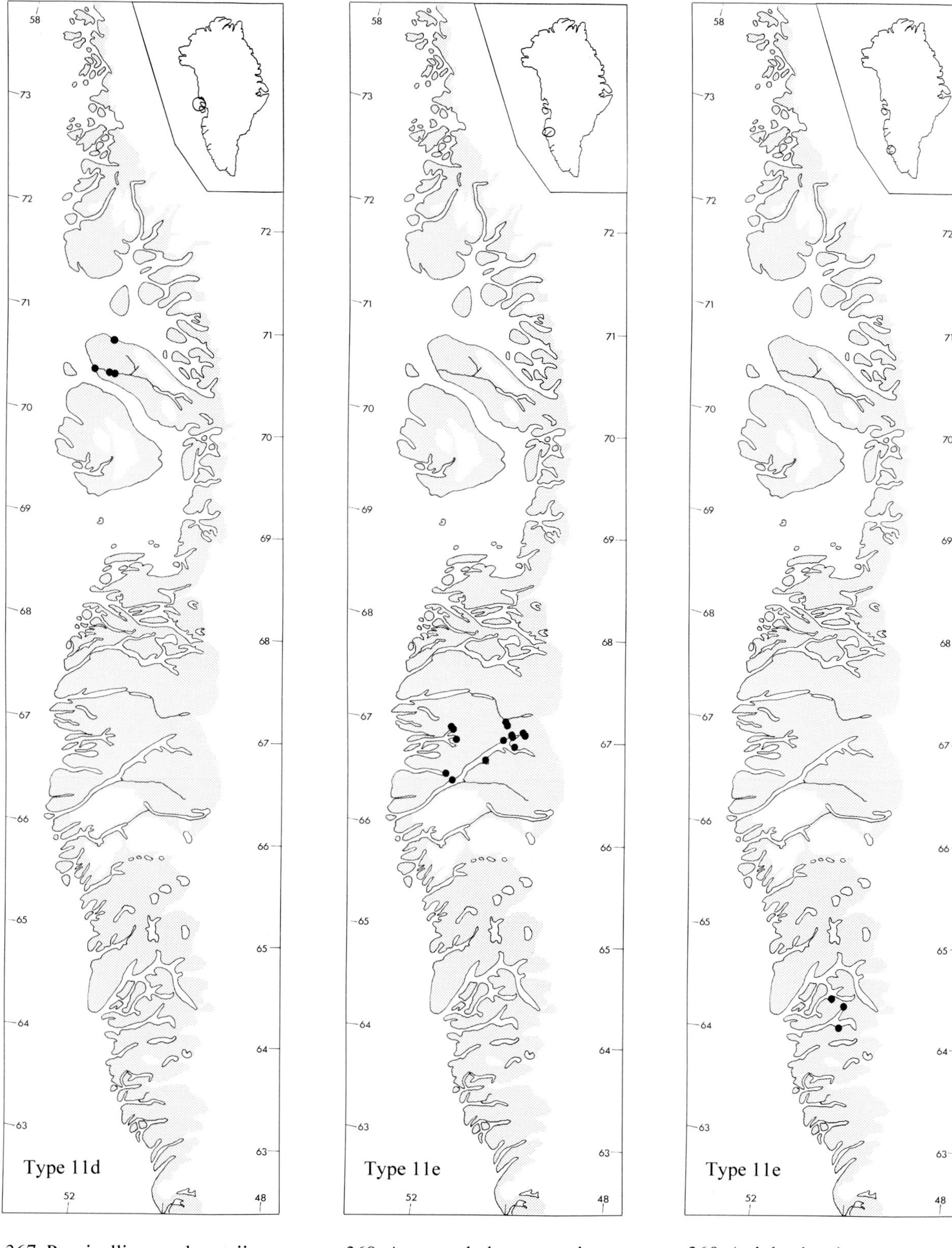

367. Puccinellia rosenkrantzii

368. Arctostaphylos uva-ursi ssp. coactilis

369. Atriplex longipes ssp. praecox

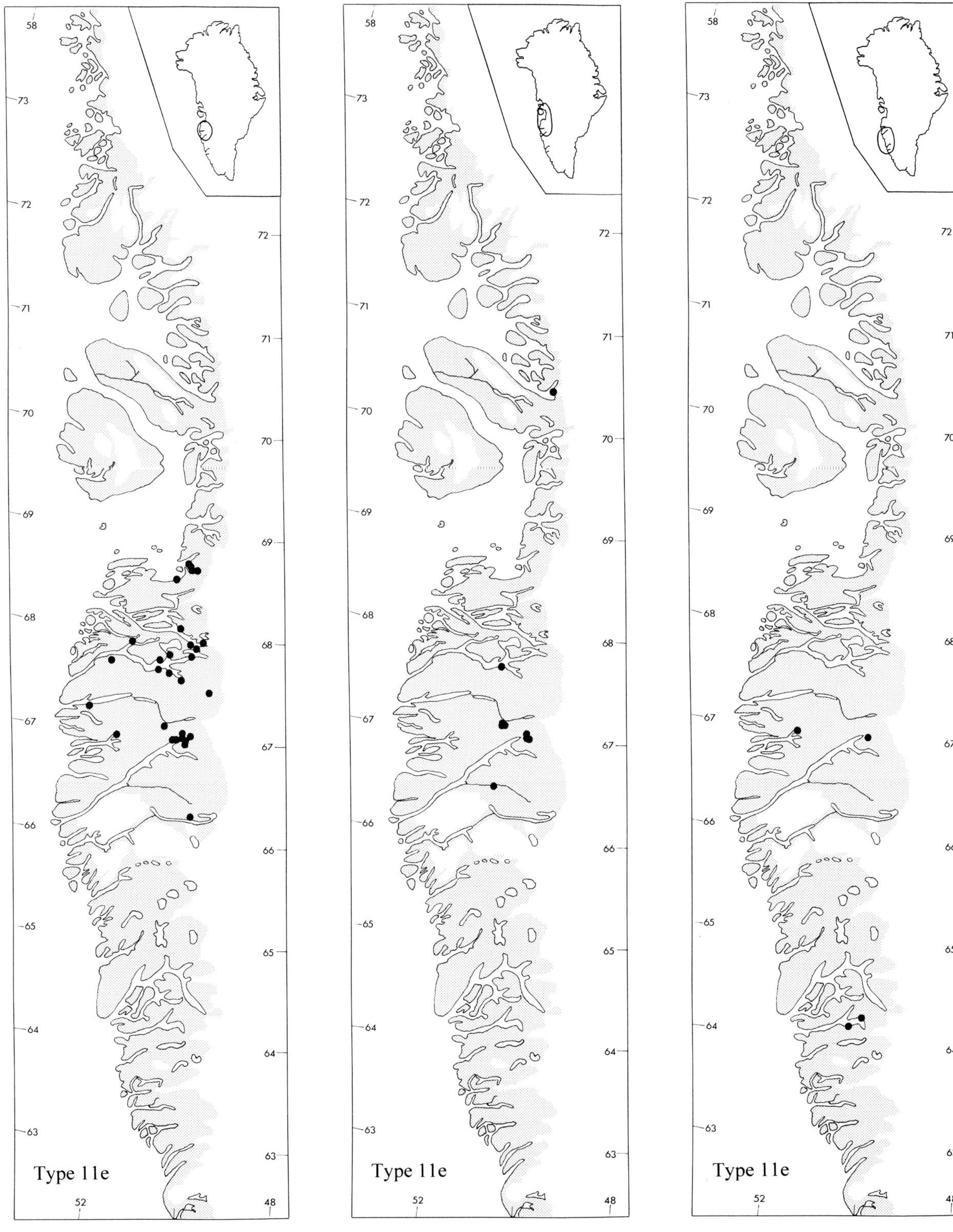

370. Calamagrostis lapponica var. groenlandica

371. x Ledodendron vanhoeffeni

372. Ranunculus cymbalaria

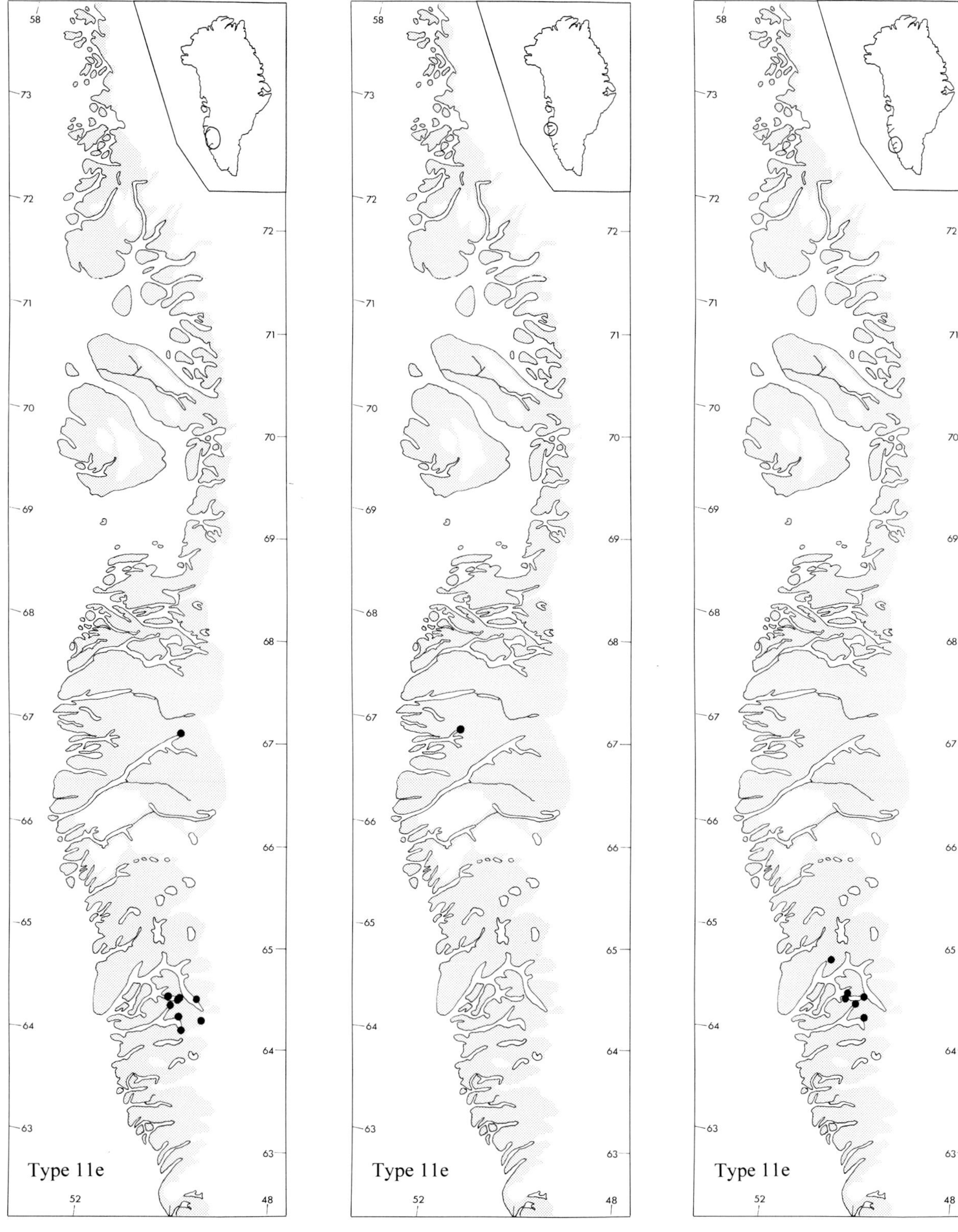

373. Sisyrinchium groenlandicum

374. Spergularia canadensis

375. Zostera marina

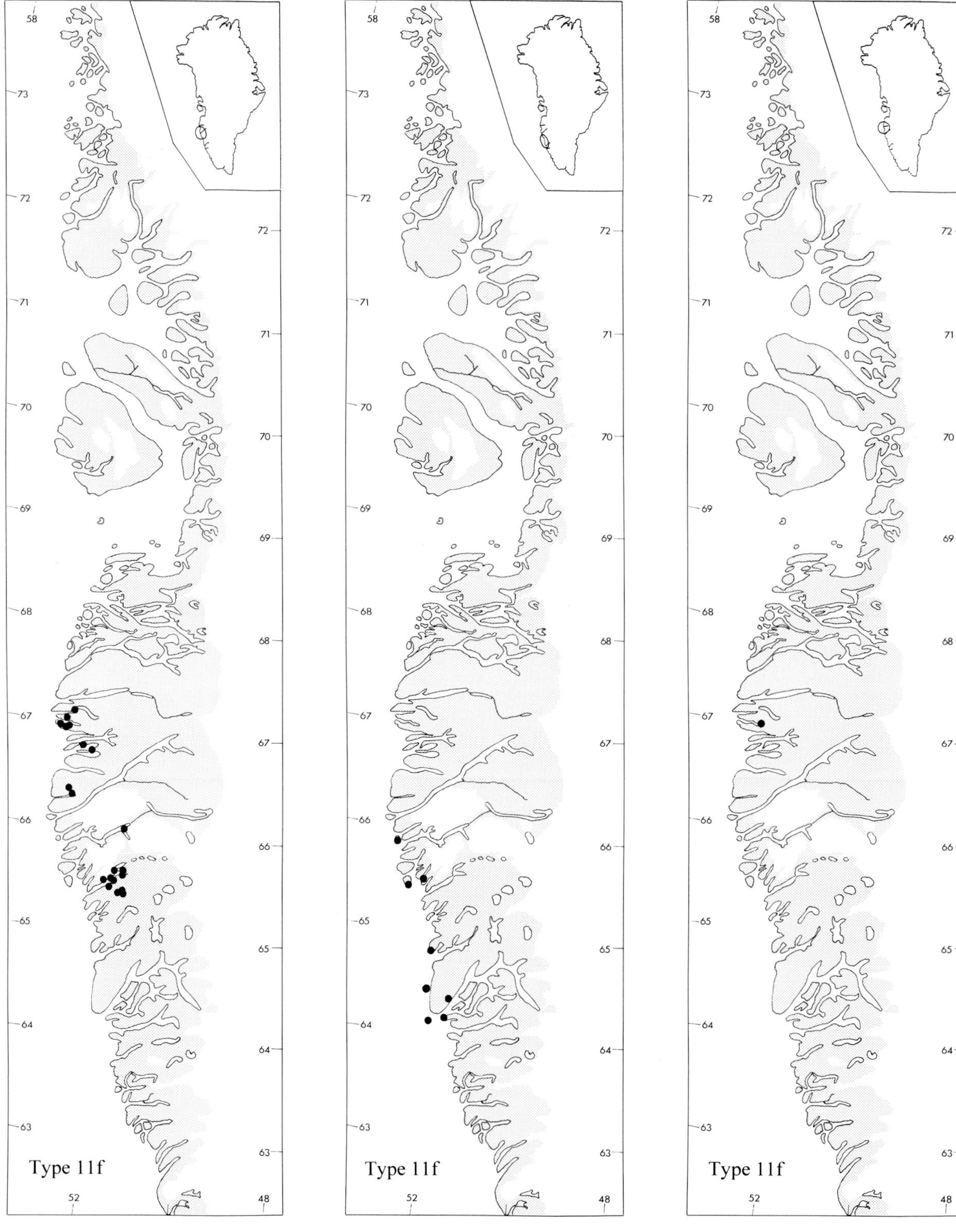

376. Anemone richardsonii

377. Arctophila fulva

378. Cerastium arvense

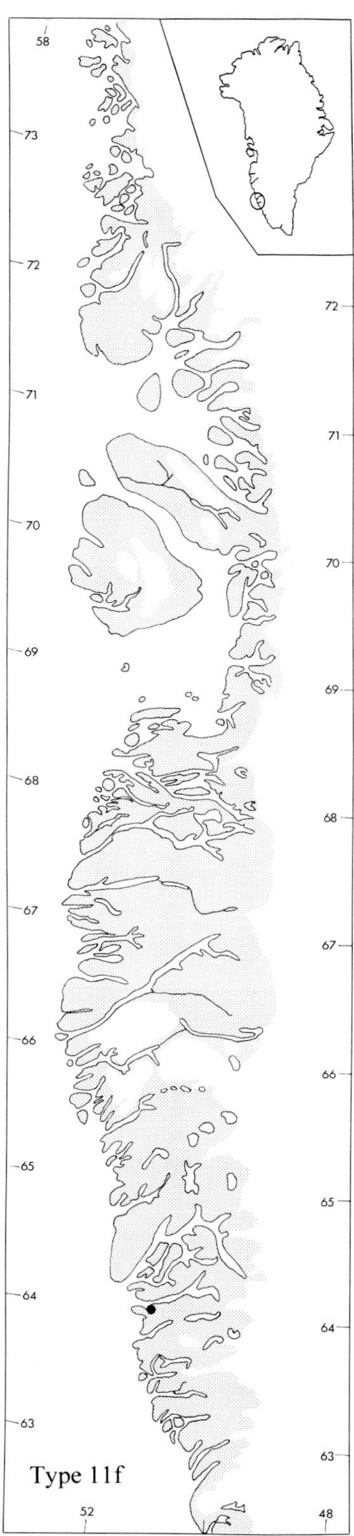

379. Pedicularis groenlandica